CONVERSING
WITH THE
PLANETS

CONVERSING WITH THE PLANETS

HOW SCIENCE AND MYTH INVENTED THE COSMOS

ANTHONY AVENI

KODANSHA INTERNATIONAL
New York • Tokyo • London

Kodansha America, Inc.
114 Fifth Avenue, New York, New York 10011, U.S.A.

Kodansha International Ltd.
17-14 Otowa 1-chome, Bunkyo-ku, Tokyo 112, Japan

Published in 1994 by Kodansha America, Inc.
by arrangement with Times Books, a division of Random House, Inc.
This is a Kodansha Globe Book.

Printed in the United States of America
94 95 96 97 98 99 7 6 5 4 3 2

Library of Congress Cataloging-in-Publication Data
Aveni, Anthony F.
    Conversing with the planets : how science and myth invented the
cosmos / Anthony Aveni.
        p.   cm. — (Kodansha globe)
    Originally published: New York : Times Books, 1992.
    Includes bibliographical references and index.
    ISBN 1-56836-021-5
    1. Cosmology.   2. Cosmology, Ancient.   3. Cosmology, Babylonian.
    4. Science fiction.   5. Anthropology.   I. Title.   II. Series.
QB981.A99   1994
398'.362—dc20   94-9211
                                                                    CIP

Text design by Anistatia R. Miller

The cover was printed by Phoenix Color Corporation,
Hagerstown, Maryland

Printed and bound by Arcata Graphics,
Fairfield, Pennsylvania

For Lorraine

CONTENTS

ACKNOWLEDGMENTS

To my colleagues Johanna Broda, Robert Bye, David Carrasco, Michael Closs, Robert Garland, Owen Gingerich, Joscelyn Godwin, Atsuko Hattori, Justin Kerr, Louise MacDonnell, Margaret Maurer, Jim McCoy, Ellen Peletz, Linda Schele, Barry Shain, Barbara Welther, Warren Wheeler, and Chelle Wolboldt, who have provided materials, dialogue, services, or all three, I offer my gratitude.

I praise the editorial skills of Susan M. S. Brown and Betsy Rapoport, the faith of Faith Hamlin, and the ever-constructive support of Lorraine Aveni.

PREFACE

Turn out the lights and watch the real ones in heaven—those our ancestors' imaginative minds used to mold a wonderful poetic imagery about themselves and their relation to the universe. For long ago the fingertips of humankind touched earth and sky more sensitively, and from those sensations there came a self-awareness that we could never be separate from nature. Our predecessors expressed their conscious presence in a living universe through a vivid and imaginative dialogue with its many aspects—with mountain, water, moon, and sun. They explained what the real world meant to them through their art, architecture, and the written and spoken word, and they passed their revealed truths down to succeeding generations, who accepted some eternal verities and altered others. My goal is to trace how and why their dialogue with nature has changed to become what ours is today.

I am about to explore the many and varied ways people from all ages have been attracted, like moth to flame, to a handful of bright lights called planets that move steadily, yet erratically, across the sky. Because the ancients accorded one planet special status, Venus will become my primary lens, through which I will focus upon the way diverse societies have framed their relationship with the natural world. They once called her the goddess of love, a fickle deity whose seductive behavior in evening twilight never gave a hint that she would betray her celestial lover in the stark light of dawn the morning after. They spoke and prayed to Venus. They paid close attention to all the subtle swerves in her promenade across the sky, for they knew well that each twist and turn would bear direct consequences for them: bad crops, war, pestilence, or fertile fields and peace with their neighbors. They asked her questions, and the answers she gave helped them to organize, order, and structure the universe around and within them.

We can never know how past people understood their universe—how they discovered the ways its manifold parts harmonized together—unless we make some attempt to see what they saw, to hear what they heard, to feel what they felt, to taste what they tasted, and to smell what they smelled. Understanding how nature and culture are related begins by bearing witness to the complex behavior of things and phenomena in their natural setting. But more than this, we must read what they wrote in order to understand how our predecessors mapped what they saw in the sky onto conceptual spaces unfamiliar to our modern rational minds—colorful mythologies about the gods and the afterlife, a world of zodiacs and astrological horoscopes. We shall discover that the underlying agenda of their mythology—creating order in nature by paying attention to its subtle nuances that impinge upon the senses—differs little from that of our modern science. And, as we journey among the crossroads of planetary myth and science, we will find that the common sense of one era becomes the superstition of another.

Anthony Aveni
Hamilton, New York
January 1, 1992

CONVERSING WITH THE PLANETS

THE PROCESS:

A COMMON GROUND OF DISCOVERY

Ignorance is the curse of God,
Knowledge the wing wherewith we fly to heaven.
—WILLIAM SHAKESPEARE, *2 HENRY VI*

Recently, while visiting Mexico City's National Museum of Anthropology and History with a group of friends, I spotted a sculpture I had not seen before. Tucked away in one corner of the Aztec Room was a jadeite representation of a calabash, or gourd-squash, about a foot across and carved out of a single chunk of stone. No real-life squash ever could have looked like this. The artist had rendered the mature, ready-to-eat Aztec staple with its attached flower opened to full bloom. Now even a weekend tiller of soil knows that when the flower is fully open and developed, the squash barely will have sprouted; or if the squash has fully matured and is ready to pick, its flower long since will have withered and dropped off. Clearly the native sculptor was capable of creating a beautiful work of art, but he was not a very keen observer of nature.

At least that is what I thought until I noticed another carving nearby. It was a representation of maize, that other basic Aztec commodity, executed in a similar guise. The image, carved this time in granite, was of a serpent, which symbolizes the earth's fertility, and on its back were a number of fully matured maize cobs, their husks pulled back and neatly braided, tamale style. A flowery tassel in full bloom was attached to each cob. As with the squash, the maize sculpture depicted two stages of growth that can never happen simultaneously. Each half of the calabash and maize sculptures seemed totally faithful to what I have actually seen in my garden at the beginning and the end of the season: the correct number of petals on the squash bloom, the perfectly shaped ribs on the mature squash, the corn tassel that hangs down and blows in the wind just the way it does in the field, and the precisely articulated kernels of mature maize.

But the artists who made these carvings just a few generations before Cortés landed on Mexico's shore half a millennium ago were neither naïve nor inattentive. They were only expressing knowledge of the world about them in a way that is unfamiliar to us. In each case they had conflated different stages of plant metamorphosis into a single coherent image. If this bothers us, perhaps it is only because, as Darwin's heirs, we live in a world that stresses evolution, development, and change. The story lines we create to explain nature follow the passage of all things, animate as well as inanimate, through temporal stages— quasar to galaxy, gas cloud to star, streambed to canyon, ape to human. The compound imagery housed in the museum in Mexico is as foreign to us as the head of an aged woman attached to a lithe young body. But for the Aztec craftsmen who were commissioned to express knowledge about the world of plants, it was less important to show a particular growth stage in the life of the plant. Rather, for reasons that escape us, the polychronic image—the combination of realities pulled from different time frames and brought together by the human imagination into a composite whole—seems to have held greater significance. What appears nonsensical to the eye at a moment in time suddenly crystallizes into a coherent and meaningful representation of nature.

What principles of discovery lay behind the Aztec sculptors' rendering of nature's forms? Perhaps we will never know, for the heyday of their empire has passed, and precious little survives for us to piece together. Whatever the mental process that unified these unlikely botanical images, the scientist in me is convinced by the evidence that careful observation surely was a part of it. But my more recently acquired viewpoint as an anthropologist will not allow me to overlook the cultural and historical context in which the Aztecs made their discoveries. One of my goals in this book is to persuade the reader that knowing the era and context in which a discovery is made is the only way of truly understanding the process of how people acquire a knowledge of the world about them.

I believe that what is true of plants is also true of animals, mountains, streams—even stars. For example, the Maya Indians of Yucatán have always connected the planet Venus with their god of rain. It is not obvious why. The association seems incongruent. Most modern sky watchers discern no apparent repeatable aspects of Venus that follow a seasonal cycle. True, five Venus cycles (each 584 days long) equal eight years, but the rains fluctuate drastically in any single year, so how can Venus's synodic period* predict anything? But before we judge Mayan

---

* A synodic period is the time it takes a planet to get back to the same place in the sky relative to the sun.

4

weather watchers to be only casual observers of Venus, the way I almost dismissed the Aztec sculptors, we should look a little more carefully at the phenomenon of Venus's appearance and disappearance from view in the morning and evening skies.

In the Venus Table in the Dresden Codex, one of the ancient Mayan priestly books I will dissect in Chapter 4, the scribe depicts cycles of celestial phenomena that have no real analogy in our astronomy today, including when Venus is visible in the evening and predawn sky. Such intervals were canonized on each page of this table. When we chart Venus's disappearance periods over several seasonal years as seen from Mayan territory, we discover that the planet's absence is shortest when Venus vanishes in the dry season and longest during the time of rain. In other words, *how long* Venus is out of sight in the sky is a good index of *when* the wet and dry periods happen in the seasonal cycle. Such practical knowledge is very important for agriculture. The only hitch is that Venus does not appear or disappear at a fixed time every seasonal year, say, the first week of March or the last week of April.

How many modern astronomers are attuned enough to Venus's movements to recognize that connection? Because we see what we are trained to see, we tend to overlook seasonal correlations that do not occur at a fixed point in the year every year. Like those ripe Aztec vegetables with flowers in full bloom, the Venus Table weaves a composite image out of a host of Venusian* aspects that can be seen at different times, an image that reveals how the planet is connected in a very complex way with the rainy season.

Let me cite another example of the difficulty of focusing a problem through the correct cultural lens. In the fifth century B.C., Plato posited an imaginative structure for planetary orbits; he described them as concentric whorls on a cosmic spindle, the axis of which was spun by the Fates—goddesses who presided over the lives of humans. Just how fast and which way the parts of the spindle spun were matters of deep concern to Plato. Among the remotest planets, defined as those that moved with the slowest speed on their orbs about the earth, Plato carefully singled out Mars because of its counterrevolutionary motion.

Early astronomers had observed counterrevolutionary or retrograde motion as an abrupt halt and temporary reversal of direction in the normal course of a planet across the background of constellations—a warp in the fabric of planetary time. Jupiter and Saturn execute one such loop in each of the years it takes them to make a full cycle around the sky. Thus, Jupiter traverses the sea of stars in twelve years, making

---

* British science writer Patrick Moore, in his popular book on modern-day Venus, writes: "There is no generally acceptable adjective for Venus. 'Venusian' is common but ugly; 'Venerean' is even worse. 'Cytherean,' an adjective derived from the old Sicilian name for Venus, is perhaps preferable, though not strictly correct." Moore uses the last. I will employ the first.

twelve retrograde loops, whereas Saturn completes twenty-nine loops in its twenty-nine-year cycle. But Mars makes *more* than one full traverse over the starry background in the time between its retrograde cycles. The Greeks were transfixed by such counterrevolutionary phenomena, which stood in the way of formulating simple earth-centered models of the universe. To the eye trained to spot retrograde loops, Mars's way of marking time seems clearly out of joint when compared with those of its more lethargic brethren.

My tales of Venus and Mars, squash and corn, are designed to entice the reader away from some of the widely held notions about the discovery and exploration of nature that are ingrained in modern culture. I want us to walk some worthwhile paths that are not so well trod. The example of Venus and the rain demonstrates that associations among natural phenomena that appear to have no connection with one another can be revealed by sharp-eyed observers and that they often have meaning in a foreign cultural context. The botanical and celestial parallels may carry us far from our own sphere of common sense, but these examples show that nature's revealed truths need not be the same to all eyes and minds at all times. In these pages I celebrate this diversity.

Understanding someone else's viewpoint can be difficult. Human expression is complicated by differences in language, education, and training as well as general outlooks on the world. Anyone who visits another culture well off the mainstream of his or her own society becomes acutely aware of human diversity—the way others worship, eat, relate to one another. Curiously, socialized human beings do not naturally take to diversity; rather, they tend to suppress it. When we attempt to piece together and confront ideas shaped in the heads of the people of long-vanished civilizations, vestiges of knowledge that lie hidden away in symbols written in dusty old texts, hammered on clay tablets, chiseled into sculpture, or painted on wall murals, we feel even more estranged. Can we really hope to understand their ideas by looking only at the material record that survives them?

"Every age has the Stonehenge it desires and deserves," said British historian Jacquetta Hawkes in her essay "God in the Machine." She was responding to attempts by modern science to explain the mystery of Stonehenge, Great Britain's most famous ancient megalithic monument, as an astronomical computer. Our typical response to the science of the past is to people it with imaginary ancestors, cardboard cutout images of ourselves scaled down into less evolved versions of the modern quantitative scientists who plumb the depths of the universe with satellite and computer. This response is too simplistic.

A better way to get at what went on in other peoples' minds might be to suspend temporarily belief in the notion that only the knowledge we have acquired by walking the particular paths of discovery taken by our predecessors has intrinsic value. Suppose, instead, we try to formulate a much broader definition of discovery that applies to both today's scientist and yesterday's Mayan and Babylonian wise men.

Is there a common ground of discovery, a perspective that can place the ways people explain natural phenomena on a broader cultural base? Is there a trait or habit that all human beings who try to account for the natural world around them share? Mexican Nobel laureate and writer Octavio Paz has said that *imagination* lies at the basis of all discovery. Whether in the artist, poet, or scientist, this is the faculty that reveals the hidden relations among things. A gifted poet's imagination deals with feelings, a perceptive scientist's with the world of natural processes, and a brilliant historian's with the reconstruction of events of the past in ways that have never been revealed before. Imagination leads to the discovery of the secret affinities and repulsions that explain things. It makes visible that which before was invisible.

Physicist Jacob Bronowski goes further. He claims that the process of discovery for both scientist and artist is identical. It begins by contrasting two unlike appearances. Then, like a flash of light, out of that contrast emerges a hidden likeness, a revelation no one ever had posed before but one that can be shared by those who are properly indoctrinated. Whether it be the appreciation of a work of art or regard for a law of science, the process is the same.

One scientific example of this discovery process is well known to anyone who has studied science in high school. It consists of the juxtaposition of two unlikely images, an apple hanging in a tree and the silvery moon suspended from the sky. When Sir Isaac Newton witnessed the apple fall, he is supposed to have wondered whether whatever power drew it to the earth might also pull upon the moon, thus keeping it in orbit. Newton's discovery principle is the concept of gravitation. The earth's gravitation unites the apple in the garden with the pale moon in the sky. The unity is expressed in a single mathematical law that describes the movement of each. Two dissimilar apparitions joined by a universal principle that forever after establishes their underlying sameness: that they possess mass and therefore mutually attract every other object in the universe. Properly indoctrinated students of elementary physics—those who are facile with mathematics, who consider Newton's laws, and who experiment with falling bodies—can re-create in a laboratory the essence of what Newton discovered. They can share in the discovery process.

The artist, too, creates unity through likeness. Bronowski's favorite example is Leonardo da Vinci's *Lady with a Stoat*. This painting shows a young woman stroking a reddish brown ermine that she holds in her arms. The anatomical characteristics of the girl are mirrored in the beast, especially the stately, yet brutal and stupid-looking animal quality of the girl's head; the gestural postures of the girl's hand and the claw of the beast suggest a further anatomical comparison. Moreover, the animal was an emblem of Leonardo's troublesome benefactor Ludovico Sforza as well as a pun on the name of the girl who was the usurper's mistress. To those properly indoctrinated—those who know fifteenth-century Milan—Leonardo's painting is as much a disclosure of hidden likeness as Newton's universal law of gravitation is to one who knows classical physics.

Clearly then, the use of the imagination as a vehicle for scientific discovery has its parallels in the world of art, music, and poetry, and we can legitimately speak of the discovery of patterns in a Cubist scene viewed through the eyes of Picasso or the likeness between the ages of man and the seasons of the year in the mind of Shakespeare—all noble examples that excite our imagination. But what of the perception of basic similarities between animal characteristics and the human temperament as seen in the mind of a medieval wise man, or between the sinuous motions of a snake and the celestial course of Venus articulated by the Mayan priest? Why is it that, when used in these alien contexts, the words *discovery* and *imagination* seem to lose their luster? Is it because they reunite certain entities we have been taught to think of as unmixable as oil and water, because they intermingle the spheres of inquiry of science and mysticism?

One of his biographers tells a story about Johannes Kepler, the seventeenth-century German astronomer who had spent a large portion of his life using data from observation to determine the sizes and shapes of planetary orbits. Was there a single mathematical or geometrical law, he wondered, that governed a planet's distance from the sun? One day, while lecturing to his class at the University of Graz in Austria, Kepler drew this figure on the board:

Suddenly, he was struck with the idea that the placement of one geometrical figure within another might hold a key to the answer to his question. Good mathematician that Kepler was, he knew that there were only five regular polyhedra—solid figures whose faces are composed of identical polygons. (These five magic figures are the tetrahedron—composed of four equilateral triangles; the cube—six squares; the octahedron—eight equilateral triangles; the dodecahedron—twelve pentagons; and the icosahedron—twenty equilateral triangles.)

A famous geometrical proof demonstrates an essential quality of regular polyhedra: Spheres can be inscribed within each regular polyhedron such that they touch the center of each face of the polyhedron. Also, spheres can be circumscribed about these figures such that the corners of the polyhedra touch the spheres. Now, Kepler also realized that there were six planets orbiting the sun (this was, obviously, before the discovery of Uranus, Neptune, and Pluto) and, consequently, five spaces between them. Was this a coincidence, or had God deliberately designed the architecture of the universe so that the five regular polyhedra, each in its correct place, would fit exactly between the planets' orbits around the sun? Kepler is said at that moment of revelation to have dropped his chalk, fled the classroom, and sequestered himself for an intense, lengthy encounter between the axioms of God-given solid geometry and the dynamics of planetary orbits. Convinced he was on the right track, Kepler even spent a large portion of his salary to construct a model of the spheres and polyhedra that fit one inside the other, like "monkeys in a barrel."

Was Kepler mad? Actually, this has been a favorite subject of psychologists interested in the relationship between creative genius and schizophrenia. Aberrations seem to have run in his family. His mother had been accused of consorting with the Devil, his great-aunt was burned at the stake for practicing witchcraft, and several of his siblings died at an early age from various maladies. He himself was sickly, with bad eyes, skin disorders, worms, hemophilia, digestion problems, even hemorrhoids. Bullied by his father, ignored by his mother, unpopular with schoolmates, it is a wonder enough of his character even survived the ordeal of life, much less created patterns of order in the solar system that yet have value.

As Kepler himself would later be forced to admit, his polyhedra theory was wrong. And yet, as foolish as the whole affair looks to us, the process of finding unity between polyhedra and planets exhibits the same potential quality of imagination as Newton's apple and the moon.

"There is geometry in the humming of the strings. There is music in the space of the spheres." Kepler was very much influenced by Pythagoras's words, written more than a millennium before him. He took them to mean that God's secret was encoded in a series of planetary musical tones. Equating planetary speed with musical pitch, Kepler believed that the faster planets trilled out high notes while the slower ones growled in the bass register. Together they sounded a heavenly symphony ordained by the Creator. When he attempted to write out God's musical score, Kepler happened upon his harmonic law—the one that mathematically relates a planet's period of revolution about the sun to its distance from the sun. This law was one of the keys to Newton's brilliant discovery of the law of universal gravitation, which we still employ to determine how long a Voyager or Magellan space probe takes to get to its planetary destination.

Today we credit Newton with genius for having made a discovery, but we discredit Kepler for having followed the lead of a nonsensical revelation he experienced in a classroom. But in the Europe of Kepler's era, it would not have been unreasonable to think of God as a universal craftsman or even a divine musical composer who set the planets into motion each with its particular pitch to create the Harmony of the Worlds.

Mine is not a book about silly ideas, but it is dedicated to exploring the context—the cultural as well as the natural environment—in which a variety of explanations about the behavior of the planets have been framed. Whether we find revealed likenesses that are still valid is less important to me. More important is that the mental process of scientific discovery has not changed. We create order and unity by bringing together seemingly unrelated phenomena and concepts. If we look only at whether the results of science are right or wrong in an absolute, nonhistorical sense, we run the risk of believing that any ways other than our own of understanding and explaining nature have no value, that they never were part of structured and meaningful thinking, that they had no context worth examining. And this reduces the possibility of understanding the origin of our own modern scientific concepts and exactly how they differ from those espoused in faraway places and remote times.

To create unity is part of a natural human desire to seek order, to construct a world less fraught with dissimilitude. It is reasonable to assume that, of all nature's events, those that happen in the sky, because they are the most dependable and reliable our senses confront, would offer the ideal role model to which organized societies would turn to seek structure in themselves, to discover hidden patterns of behavior between their lives and the lives of the stars. Astrology as well as astronomy began when people realized that celestial periodicities offered the ideal numbered blank pages on which civilization could write its history.

The rains might come late and summer be excessively dry; animals might migrate earlier than anticipated, and a season could well pass without berries on the bush. But the sun always appears on schedule, the moon is always full at monthly intervals. Stars, like ideal people, never deviate from their courses, and the planets will execute their loops and turns in the future precisely the way they have in the past.

No wonder astronomy is our oldest science. Its practice is regarded as a hallmark of intellectual attainment among highly stratified societies: Pharaonic Egypt, with its star-ceiling sarcophagi; Classical Greece and its detailed calendrical prescriptions for festivals and complex theoretical models of the solar system; China, with its gear-wheeled planispheres that charted the course of the emperor's celestial source of power. All around the world, heaven's acts emerge as the supreme archetype of precise harmony and metronomic repetition, from the detailed daily and seasonal scheduling of ritual to the very essence and source of royal authority. Knowing the sky has always been important.

Everything we learn about the sky today we acquire from reading books and maybe paying a visit to the planetarium. Except perhaps when we take out the evening trash or walk from the commuter train to the car or from the car to the house, casting an upward glance to see whether it might rain tomorrow, we live in a world mainly unaware of the one-half of visible space that lies above eye level. But what do we need to know about the sky in order to get into the skin of ancient astronomer-astrologers—to eavesdrop on their conversations with celestial deities? To begin with, we must temporarily divorce ourselves from the contemporary planetary imagery embossed on our minds by traditional learning. We must forget those indifferent, nonconscious worlds that move with blinding speed in elliptical orbits about a middle-sized yellow-hot star held fast by impersonal mathematical dictates. See Venus and Mars instead as the ancients saw them, and you begin to appreciate that, far from being fear-racked, backward individuals who cringed beneath a sullen sky, handicapped by never having tasted the fruits of modern science and technology, our predecessors were in a real sense more aware of the subtle essences of land, sea, and sky, more directly in contact with the world around them and the way its parts harmonized, and far more imaginative and expressive in their outlook toward it than most of us.

Even though trained as an astronomer, it took me a long time to appreciate that all unaided eyes do not acquire the same imagery when they focus on starlight. I had worked with the big telescopes at the Kitt Peak National Observatory in Arizona and taught astronomy at Colgate for more than a decade before I became interested in the possibility that each of the two hundred generations of astronomers who lived before me might have looked at the same stars I saw and seen a different light. I learned that I live on top of an imaginary chronological pyramid of progressively acquired scientific truth. I never needed to penetrate the support structure of the layers beneath me that made my truths real. It took a trip to see Mexico's past to turn my attention to the first of a host of unopened astronomy books of other societies and to shift the focus of my scientific inquiry to anthropology, for it is within the study of human culture that we must place the discoveries of all astronomy, past and even present.

I remember that trip well, for I and my students were detained by officials for prowling around on Teotihuacán's sun pyramid. We thought we had sufficient permission to climb up top at night to measure the pyramid's orientation with a surveyor's transit. My students had talked me into organizing the trip as part of Colgate's off-campus January term to investigate the astronomical orientations of pyramids, a popular idea in the late sixties, in the aftermath of the great Stonehenge controversy. (Escaping the vagaries of the harsh winter environment of upstate New York provided another motive.)

The whole story is not as important as the punch line, which came three years later, when I realized that the best explanation for the sun pyramid's odd skew was to point to a celestial event that was used to mark the start of the celebration of the new year. This occurred on the day the sun passed the zenith or overhead point in the sky—an event that can never occur outside a band around the world that lies between the Tropics of Cancer and Capricorn. We had uncovered an astronomical calendar with a New Year's Day that had no determinable counterpart in our Western system of astronomy. And the clue to the orientation puzzle emanated from a study of the written documents that came down to us from just after the Spanish Conquest, documents that detailed how and when the Aztecs worshiped the stars. Zenith sun watching is part of a three-thousand-year-old tradition in ancient Mexico. So that trip was my first encounter with a sky I had never seen before but one I would need to study carefully in order to understand my own ancestors as well as those of modern-day Mexicans.

In Chapter 2, "The Images: Planets and Sky," I am going to explore the naked-eye sky—the way the peasant in the field or the king on his throne saw Venus twist and turn as it passes back and forth across the blinding solar light from month to month and year to year, how Mars slows and suddenly halts in its zodiacal course, then turns round and goes backward awhile. We will see the planets rising and setting at different points on the horizon, disappearing in the light of the sun for lengthy intervals that can be tied to a host of natural events—the timing of the seasons, the direction of the wind, the coming of the rain. The value of human knowledge changes with time and place, and much of this old planetary stuff has been forgotten. That it has become of little direct concern to us now is our loss, for it forms an integral part of the story of how people recognized, categorized, and explained what they observed in nature, how these explanations changed throughout human history, and what made them change.

Venus will be a special focus of both this chapter and the entire book because in many of the ancient stories it has been singled out and accorded special status. Not only is it so very bright but also it behaves in a way noticeably different from that of the other planets, so different in fact that it will force us to question whether it really makes sense for all civilizations on earth that paid attention to the sky to classify all the wandering lights under the single heading *planet*. And this raises the question of why Western astronomy has chosen to do so. We need to examine carefully how Venus's bright white light shifts and flickers, comes and goes, brightens and fades, cycle upon cycle, and how Venus's aspects contrast with those of the other cast of characters that have paraded across the celestial stage since long before a human audience ever assembled to watch them perform—the sun and the moon, Mars, Jupiter, and Saturn. (The remaining planets have been known to us only since the advent of the telescope.)

Learning what happens in the sky as perceived by the naked eye will set the stage for a series of questions about sky watchers all over the world. How did people use these celestial images to create the myths and metaphors we read in their surviving literature and art? What set the standards of truth for them? What made their truths become falsehoods for us? In the intersecting world of science and mythology—astronomy and astrology—that I am about to explore, knowledge was often organized and categorized in unusual ways.

Is the universe already there—a single entity filled with matter and radiation both seen and unseen and governed by a fixed and everlasting set of rules waiting out there for us to discover, or are there infinite ways to piece the cosmos together? This question may slightly oversimplify what I mean by subtitling this book "How Science and Myth Invented the Cosmos," but it does reveal my own answer. Scientists usually opt for the first alternative. They seem united in their goal of uncovering nature's perceived underlying truths. But mythologists offer us another alternative. Placing a far heavier burden on the human mind, they suggest that there may be many answers, each valid in an appropriately understood framework, to the question How does nature work? I am about to share some of those answers by exploring the manifold ways Greek and Mayan, Babylonian and Aztec people discovered relationships between things they saw in the sky and events that happened in the world of everyday life.

Chapter 3, "Mythology: Naming the Images," is about the stories people have made up about the planets. I hope it gives a feel for the discovery process that has revealed in so many instances the hidden unity between planets and people—people who asked different questions and often sought different information to acquire answers. Why did Venus, rather than Jupiter or Mars, become the fickle goddess of love in the Old World and a resurrected hero with five faces in the New? You will see that the answer lies not so much in whim or fancy, not in the irrational superstition we usually assign to myth, but more often in what careful planet watchers perceived in those bends, kinks, and turns in Venus's journey across the sky that I will have mapped out in Chapter 2. A change of color, subtle fluctuation in brightness, the moment a planet stood still or stretched to its visible limit from the sun, when it first began to descend or when it convened with one of its sisterly denizens—all these phenomena evoked specific messages in the broader planetary dialogue conducted by direct word of mouth between celestial gods and the worshipers who spoke their names.

It is humbling, perhaps even worrisome, to think that ancient people, with minimal technical apparatus, might have developed an acuity sharper than ours to sense natural phenomena they sought to express symbolically. Can we really believe they saw patterns in nature that are not known to us? After all, they were conditioned by a lifetime of immersion in an environment of particular sights, smells, and sounds that nourished and honed their sensory appetites. And they were unencumbered by the dulling effect on the senses of the technological dependence the modern world has acquired.

For at least the last two centuries (some historians will say even longer) we have been taught to think that the forebears of modern West European–based culture—the Babylonians, Egyptians, even the Greeks and medieval and Renaissance societies of West Europe—were intellectual lightweights, lacking in many skills we now possess. Our focus is on the present. That is why we often portray our ancestors as misguided individuals who only served the purpose of laying down the building blocks that support the modern ladder of progress. We respect them perhaps not for what they were but because of what they did for us. Because of our cultural astigmatism, Einstein means more to us than Newton, and we praise Galileo more than Aristotle. The closer one of our intellectual heroes is to us on history's time line, the greater our admiration. We have a hard time taking ancient systems of thought and worldviews seriously. Imbued in the dye of our progressive, contemporary fabric, we rather pity our distant ancestors for lacking the printed word, for possessing only crude and rudimentary technologies, for being incapable of wiping out the many plagues that killed them.

What about the ancestors of *other* cultures, those farther in space as well as further in time from us—the Maya of Yucatán, the Incas of Peru, the Polynesians of Oceania? We explore them to a far lesser degree even than we do the ancestors we presume to be our own. When we do, we use rather vaguely formed impressions of these people to construct a kind of comparative cultural mirror from which our own reflection usually shines forth brightly. After sketching out the origin of the universe according to a Norse myth, cosmologist Steven Weinberg subjectively summarizes: "But I think it is fair to say that this is not a very satisfying picture of the origin of the universe. Even leaving aside all the objections to hearsay evidence, the story raises as many problems as it answers."

We must face facts: In many ways our predecessors were *not* like us. When it came to categorizing natural phenomena, they were integrators, not differentiators. They did not compartmentalize their knowledge of the world into separate, unrelated fields of study—such as biology, geology, chemistry, and physics—and they did not reduce events to chain-link cause-and-effect explanations. Most of them exhibited a tendency to interrelate the behavior of things in the sky with other happenings in nature that they captured in their universe of senses. For example, the Aztecs saw blood tied to the sun, the Maya connected the movement of bees to the course of Venus, the Egyptians linked the shape of a ball rolled around by a dung beetle to the roundness of the sun on its celestial journey—these likenesses make no sense to us but they had far-reaching implications for their discoverers.

They painted their knowledge of the natural world on a far larger canvas than we. The sky myths they created joined a world we regard as inanimate to the animate sphere of their own lives, the unfolding of their history, politics, social relations, ideas about creation and life after death. They forged links between the sky and just about every phase and component of human activity—what we call astrology. And they celebrated this knowledge not only in text but also in art, architecture and sculpture, poetry and song. Much of this knowledge was sophisticated and highly organized. We deserve to appreciate it, to be enlightened by the reflection of a very different comprehension of nature.

Chapter 4, "Astronomy: Following the Images," is designed to demonstrate that even though it may deal with a different imagery and be directed toward different ends, ancient science was often paradoxically more like its modern counterpart than most of us think, for it too could be rigorous as well as capable of yielding precise predictions. This chapter, necessarily a bit more technical than the others, probes the elegant mathematical intricacy of two Venus prediction tables from opposite sides of the world. One is inscribed in cuneiform writing on clay tables of the seventeenth century B.C., dug out of the sands of the Fertile Crescent. The other, written in Mayan dot-bar notation on lime-coated bark paper books which surfaced in nineteenth-century Europe, was originally composed in the jungles of Yucatán three centuries before the Spanish Conquest. The cultures that spawned these texts were very remote from each other in both space and time. Yet these documents, looked at side by side, teach us that religious belief can both constrain and modify the way people express their scientific records. Here was planet watching elevated to its most quantitatively complex form—elaborate, precise, and rich in detail. It may not be real social history but it is real astronomy in every sense of the word. But the texts have something else in common: Basically they tell the same story—they sing the same song, almost note for note. Each describes the planet Venus marching to a lunar beat, a rhythm quite foreign to our modern way of marking time. From the primacy of the day to the length of our year, we follow the sun rather than the moon as time's true measure. I will explore why.

Our ancestors did more than provide a springboard from which we eventually developed the modern scientific understanding of the universe. When we decode the messages they have left behind, we hear sounds that ring dissonantly in our ears—the unfamiliar Venus of occult astrology, the exotic Venus bearing omens, the omnipotent Venus bestowing the right to rule on kings.

We call our modern celestial endeavors scientific astronomy, but we label as astrology most ancient and foreign forays into the sky. Some of us might come away disappointed when we are forced to accept that the message revealed by the two Venus texts in Chapter 4 is partly divinatory—concerned with celestial prognostications about human behavior. But we should realize that by totally separating ancient science from mysticism we only distance ourselves from our predecessors. Such a division helps little in understanding their ways of thinking.

In Chapter 5, "Astrology: Believing in the Images," I raise a broader set of questions based on the previous chapter: What *is* planetary astrology? How was it intended to work? What did those societies that developed it have in common? Under what conditions do people create systems of belief that operate on the principle that messages which come from the sky are intended for them? Why does the modern scientific community cast a glaring eye of suspicion upon astrology? Is this a paranoid rejection, and, if so, what does it say about us? Devoid of its ancient cultural underpinnings, today's version of astrology emerges as a misfit ideology, one that deceives modern practitioners who attempt to lift it out of its once-appropriate human context and fit it into a radically different worldview.

In ancient America, as well as in China and India, the proliferation of complex astrology was accompanied by the institution of the belief that the ruler possessed supernatural powers—that royalty was celestial power incarnate. The surviving record suggests that this belief developed only when society had become highly stratified. For the Classical Maya, the continuity of the power of rulership was directly expressed from the sky-creator in the form of discernible subtleties in observable planetary cycles—Venus most prominent among them. In the Old World of medieval times, the planets were believed to possess direct ties with the human body and with concepts of healing, a notion expressed particularly in early Renaissance art. Astrological symbols of the planets can be traced all the way back to the pagan mysteries of the ancient Classical Mediterranean world, which later became incorporated into the ideology of the Christian church. In all cases, these representations were not devised irrationally or thoughtlessly. They were part of a system, a total scheme for understanding nature that is worthy of our attention.

Our distaste for much of ancient astrology's imagery stems from the negative reaction of science to the feeble remnant of ancient astrology—a system whose basic mechanics we must explore, if only to demonstrate exactly how its beliefs do not square with today's cultural norms. The Renaissance examples of planetary astrology reveal a culture on the verge of ideological overturn, about to trade a long-held belief in the direct link between people and nature for a principle of detached reason that had lain dormant and unseasoned since Classical antiquity. It is here that modern science, as we know it, really begins.

In Chapter 6, "Technology: Harnessing the Images," I follow some aspects of the great schism between God and nature that developed during the Renaissance by showing how two of Europe's first experimental scientists used the earliest form of the telescope to probe the planet Venus. The lesser known, Jeremiah Horrocks, remained firmly attached to the church's interpretation of the place of God in the natural world, a foundation that was in fact trembling beneath his seventeenth-century feet. The other, the celebrated Galileo, appears instead to have been negotiating the great ideological trade-off: the separation of the actions of humankind from those of nature. His description and interpretation of his own observations sheds a different light from that of Horrocks, one that better illuminates the path down which we have trod the last few steps out of the Renaissance into the modern age of science. For the first time, the planets were conceived truly as worlds other than our own.

I entitle the final chapter "Science: The Image for Its Own Sake," which correctly expresses the way modern science comprehends the universe—as an entity that functions on its own, without regard to our presence in it. Today we seek a different kind of knowledge from the planets, and we use it to different ends—not to know ourselves but to discover what we believe are the inviolate laws of nature, in which most of us are convinced the human spirit has no part. Why has the dialogue between people and planets changed so profoundly?

Unlike her ancient counterpart, the modern Venus is no lady; instead, she moves about a vaster sky, an inanimate object to be entered, and dissected with Venera and Magellan spacecraft. Space-age Venus, probed from afar and experienced indirectly through television imagery, emerges as a world like our own with continents and an atmosphere, yet an unimaginably hostile place, a fiery hell pelted by acid rain. Modern Mars is no longer her celestial lover but instead a dead world of desiccated, crater-pocked plains, ancient canyons, and empty valleys. If we consider these hollow reports a kind of dialogue with the planets, it is a far cry from the one in which people gazed at Mars and Venus and spoke their names to invoke their powers.

At the end of my final chapter, I explore some recent attempts to rejoin lost links between culture and nature. Such efforts seem to have emerged out of dissatisfaction with the scope and methods of modern science. For example, James Lovelock's Gaia hypothesis imagines our planet to be a living world in which human development and the evolution of the natural environment are coupled by inseparable processes. Applied to the central object of our affection, I ask whether Venus emerges as a test laboratory, a now-dead world we must probe to achieve a better understanding of how Gaia functions here at home. Or is it a world in whose planetary life we still partake? These new ideas about the way the world functions have both an anthropocentric and a teleological ring to them—an orientation and an attitude troublesome to modern science.

To dismiss such ideas as "New Age" would be callous, for they offer one perspective on a great debate among philosophers and scientists looming in the twenty-first century that holds the potential to change the way we think about the natural world. The issue: Is modern science—objective, socially neutral science—really the only acceptable road to truth? Would any wise extraterrestrial come upon the same logical system we have to explain the structure of the universe? Yes, argues Nobel physicist Sheldon Glashow. At the other extreme, perhaps there are as many roads to perceived truth as there are species that could have evolved to think them up. Philosopher Mary Hesse retorts that scientific theory is just one way humans have tried to make sense of our world, one among manifold sets of myths, models, and metaphors. There is nothing special about it.

Western culture has changed before. Are we again on the verge of altering the way we think about the world? Are we going back to the future? The last time history's odometer turned over, when A.D. 1000 approached, Christian Europe discarded the notion of a literal Second Coming of Christ here on earth and began to reinterpret Holy Scripture. Today, as we conclude the second millennium, fairly credible members of the scientific community have begun to question whether the way the sciences have dealt with nature—owning it, progressively dominating and controlling it, seeking hidden, genderless laws by which we claim it operates—is wrongheaded.

I lean toward Hesse, away from Glashow, for if my dual pursuits in astronomy and anthropology have together given me any special insight, it is that any intelligent being who pays close attention both to culture and to nature cannot fail to see that the two are bound together. My traverse of the sky will demonstrate that the way people live has profoundly affected the way they create their understanding of the natural world.

The epigraph to this chapter is concerned with dispelling ignorance about the many dimensions of knowledge. The words were spoken by the Lord Say during Cade's fifteenth-century rebellion in response to accusations by insurgents that he had corrupted the young and caused people to change the way they did things (he had constructed a paper mill, opened a publishing establishment, and built a grammar school). But, far from being corrupt or revolutionary, my desire is to cast scientific thinking into a slightly skewed, more socially based, critical perspective, not to put it on a pedestal and praise its works and deeds. Broadening the notion of scientific discovery can dispel some of the misconceptions we possess about our ancient predecessors. We need to see them fully clothed, rather than cast away those of their garments that do not appeal to our intellectual fashion palates.

Earlier I suggested that what is common to our senses need not be common to others. Sky phenomena that, because of our cultural and environmental bias, we tend to regard as the same the world over might have proffered little interest or applicability in somebody else's world-view. I believe it is safer to assume that astronomy is not a cultural universal—not a single body of knowledge with a set of processes that can become known to all of us who observe the sky. Let this be our guiding assumption as we set out. I begin by getting under the ancients' umbrella with a look at the skies through naked eyes.

CHAPTER 2

THE IMAGES:

PLANETS AND SKY

The mingled influences of the stars can be understood
by no one who has not previously acquired knowledge of the combinations and varieties in nature.
PTOLEMY, *CENTILOQUY*, QUOTED IN McCAFFERY, P. 77

Modern astronomers direct their telescopes upward, but
ancient watchers of the sky were far more likely to have
set their eyes, unaided by technology, upon the horizon.
Because celestial events were tied to ritual practice, the
ancients might have been more concerned with how
high one of their sky deities stood above a temple dedicated to its worship than with unanchored constellations wheeling
above their heads.

But try to comprehend a natural event without presupposing the
perspectives and patterns of organization modern science has taught us.
Who can look at the moon and not think of it as a place where astro-
nauts have walked? Or stars that are not blazing nuclear infernos at vast
distances, or Venus not as a spherical body that, like earth, orbits a
fixed sun? Because modern astronomers seek deeper causes that they
believe underlie nature's superficial effects, they explain the celestial
motions we witness directly by saying that what we see is only "apparent" to those of us who watch the sky from a peculiar vantage
point—the earth—that we assume to be fixed. In this chapter we will
be tempted constantly to ask: But what is really happening to make
things move this way? This is a temptation that we must try to resist,
although we will occasionally yield to it, because my goal is not to give
explanations in that familiar sun-centered orbital framework but instead to explore pathways to somebody else's version of truth—whether
we believe it or not.

So I begin this lesson in sky watching, not at the observatory's telescope but out in an open field in clear air, under a pitch black sky filled with stars and limited only by a distant horizon. If you just stand and stare for ten to fifteen minutes, you will begin to see the stars move. The moon, sun, and planets move in the same way, executing a full cycle of motion all the way about the sky in a day. But how you see them move depends quite strongly upon where on earth you live as Figure 2-1 shows. If you happen to be situated in the middle latitudes (say 30° to 50° north or south), you will see stars in the east glide upward and off to one side while those in the west pass downward and toward the opposite side. Stars in the northern section of the sky circumnavigate the immovable North Star, moving round and round like the tips of so many hands on a clock, while the southern constellations make dome-shaped arcs centered on an imaginary point far below the horizon. From tropical latitudes (within about 20° of the equator), daily motion in the sky is very different—stars move more or less straight up and over the sky dome and plunge straight down into the western horizon. Turning to the north, we still see the daily motion pivoted about the North Star, except that it lies much nearer the horizon. If you spin around and look south, you will find that southern sky movement in the tropics is much the same. In fact, there is a kind of symmetry the tropical-sky watcher can appreciate. It is as if he or she rather than a distant point on the sky's ceiling is the center of symmetry of the sky's motion.

Now return to the same place, clear night after clear night, at the same time of night, say, just after evening twilight has ended and it has become totally dark, and you will notice that the constellations will have changed positions in another way. Star patterns in the east will appear a little higher in the sky, those in the west a little lower. After a few weeks, the stars you originally saw close to the western horizon at dusk will have vanished behind the hills before the onset of darkness, already lost in the setting sun's glare. For example, Scorpio, which dips below the horizon at 8:00 P.M. in mid-November, will set at 7:00 P.M. by the beginning of December, by 6:00 P.M. at midmonth, and by 5:00 P.M. at the start of the new year. But if you turn around and look to the east, you will discover that replacements have arrived, rising above the horizon, stellar groups you might remember having seen there last year at this time and the year before, and the year before that.

If you return to that open field before dark and watch the sun go down, you might also note which constellation of the zodiac stands over the position where the sun disappears. In the middle latitudes in October, you will see Scorpio mark the spot; in November, Sagittarius will have nudged it aside. In December, when the sun's light swallows up Sagittarius, Capricornus moves in to take its place, then Aquarius in January. You will also notice that the point on the horizon where the sun sets shifts from night to night, like a pendulum on a clock, moving

Looking East in the Tropics

Looking West in the Tropics

FIGURE 2-1a

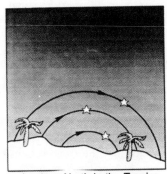

Looking North in the Tropics

Looking South in the Tropics

FIGURE 2-1. MOVING LIGHTS AND MENTAL IMAGES

What does the daily motion of celestial objects look like as seen from different places in the world by an observer who looks in different directions? In the low latitudes of the tropics stars rise more or less straight up in the east and plunge straight down in the west. In the middle latitudes the daily path lies at a lower angle to the horizon. (Compare Figures 2-1a and b, upper pairs.) Looking north or south, a tropical observer sees similiar daily motions. All the stars have circular tracks pivoted about fixed points close to the

to the right or north in winter and spring, to the left or south in summer and fall. The movement of the sunset point slows down and approaches a well-defined southerly limit as the first day of winter approaches; conversely, it reaches its northerly limit on the first day of summer. We call the times of the year when this happens the *solstices*, which means literally "the sun stands still." The sunset point moves fastest when the sun arrives at the due west position on the first days of autumn and spring.

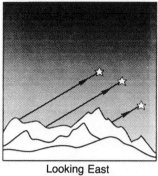

Looking East
in the Middle Latitudes

Looking West
in the Middle Latitudes

FIGURE 2-1b

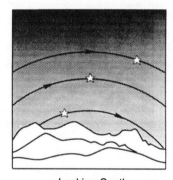

Looking North
in the Middle Latitudes

Looking South
in the Middle Latitudes

horizon (Figure 2-1a, lower pair), thus offering a sense of celestial symmetry. By contrast, an observer farther north of the equator sees stars in the northern sky whirl in twenty-four-hour circular orbits about the Pole Star, which is situated high in the sky. In the southern part of the sky this mid-latitude observer sees curved trajectories moving about an imaginary point well below the south horizon (Figure 2-1b, lower pair). (Drawing by Ellen Peletz.)

If you watch the sun rise instead of set, you will note the same progression except that north will be on the left and south on the right. Such observations marry the landscape to the sky. A careful and persistent observer could even predict where particular stars would be at a given season simply by noticing over which hill or other landmark the sun had risen or set.

The other sky disk, the one of pale silver instead of bright gold, might strike us as a somewhat more inconstant wanderer—at least it is faster and much harder to track. Like the sun and stars, the moon still rises and sets at the same tilt relative to the local horizon; however, for over half its cycle, the moon is visible for longer periods in the day than in the night sky. Sometimes for a few days it is totally absent from both the day and night skies. Another inconstancy of the moon is the way its face changes dramatically and repeatedly, from thin sickle to D shape, then to bulgy egg, and finally, like the sun, to a full, though tarnished, blotchy disk. Over the course of a number of days you could tally on your fingers and toes, you could watch the moon pass through half of its phases in the sky around dusk and the other half in reverse if you observed the heavens just before dawn.

You could also chart from memory the course of the moon among the constellations. If you imagine the starry background to be held fixed, you will notice that the moon has yet another kind of motion. It shifts about a handspan held at arm's length per night from west to east; in other words, it does what the sun does throughout the seasons, only much faster, moving opposite to the normal east-to-west daily motion of the stars across the sky. Two motions that happen at the same time can be difficult to conceive, but let me use an example that simplifies what is happening. Suppose you are in an airplane, flying west from New York to Los Angeles. You get up out of your seat and walk at a steady pace down the aisle toward the back of the plane. Your motion relative to the seated passengers back toward New York is like the motion of the sun and moon from west to east relative to the stars. However, the movement of the entire airplane, with you and all the other passengers in it, traced out along the ground by the plane's shadow, can be compared with the twenty-four-hour motion of everything in the sky from east to west.

The combined effect of these compound motions is illustrated in a month-long sequence of time-lapse drawings in Figure 2-2. If you watch the moon just after sunset from night to night, beginning with the first thin crescent low in the west, hovering above the afterglow of sunset (as shown in the position labeled 1), you discover it takes up a new position, moving that handspan per night toward the east and getting fatter, or waxing, as it widens its spacing from the sun (Days 2 and 3). After two weeks, the moon's fully illuminated disk rises in the east, opposite the place where the sun sets in the west as you stand in between them. For the rest of the month, the moon is more accessible to someone who gets up early and views the sun rising in the east, for as it takes up new positions farther eastward among the stars, you can watch it close in on the sun as its phase wanes from full to crescent (Days 26, 27, 28). Finally it vanishes in the sun's glare for a day or two as the "new moon" to complete the cycle.

The star-studded corridor that circles the sky—the one the sun follows in a year and the moon treads in about a month—is called the *zodiac*, a word that means "circle of animals." The constellations that

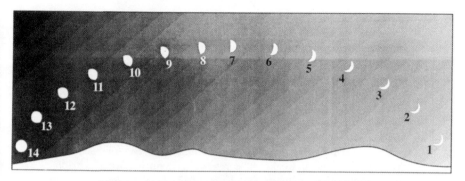

Moon on Successive Days, Looking South after Sunset

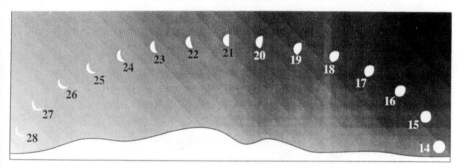

Moon on Successive Days, Looking South before Sunrise

FIGURE 2-2. THE MANY FACES OF THE MOON
The moon is the only image in the sky that regularly and visibly changes its shape over the course of its cycle as seen with the naked eye. In a month the moon moves eastward among the stars all the way around the sky, passing through all its phases as it goes. On Days 1–14, the observer looks to the south after sunset and sees the moon wax from first visible crescent to full phase. On Days 14–28, the observer watches the southern sky before sunrise and witnesses the moon wane from full to last crescent. Gone for a day or two, the moon finally returns to position 1. (Drawing by Ellen Peletz.)

mark out this band, often depicted by the figures of animals, have received much attention in legend and folklore because they chart the courses of not just the moon and the sun but also of five other bright lights—Mercury, Venus, Mars, Jupiter, and Saturn—to whose cyclic movements archaic cultures all over the world have paid rapt attention, perhaps in part because they are even more difficult than either the sun or the moon to pursue.

The Greeks called the planets "wanderers" because this handful of celestial luminaries moves against the natural grain of motion of the stars. Red Mars, bright white Venus, swift Mercury, and slower-moving Jupiter and Saturn, each possesses its own unique track across the zodiac. The Sumerians said they were errant sheep who strayed from the flock. In addition to the west-to-east motion among the stars, which they share with the sun and the moon, each planet undergoes retrograde motion. That is, in its annual course through the sky, it appears to slow down relative to the stars, to come to a stop, and for a period of several days to a few months to move backward, that is from east to west. Then it slows to a stop a second time, turns around, and resumes its normal west-to-east motion again. Figure 2-3 shows what you would see if you returned to your familiar field and followed Mars this way for several weeks. The retrograde loop is part of a triple, compound motion unique to the planets, and its recognition has had an indelible impact, at least on our Western astronomy, from the time of the Babylonians and Greeks all the way up to Copernicus and the Renaissance.

FIGURE 2-3. A REST ALONG THE HIGHWAY
Here the retrograde motion of Mars is shown in a time-lapse sequence relative to the background of stars over several weeks. First, Mars glides eastward along the zodiac like the sun and moon. But then, just as it is about to enter Gemini from Taurus, it halts, turns back to the west for a few weeks, rests again, then resumes its eastward movement to complete its loop among the stars. (Drawing by Ellen Peletz).

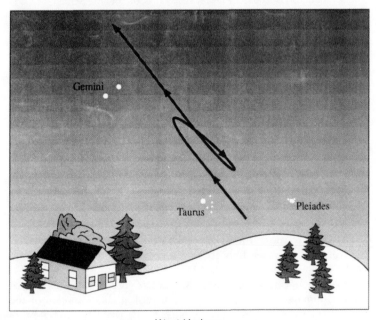

West Horizon

To extend our analogy of your movement down the aisle of the Los Angeles–bound airplane to include this additional motion, follow the course of the swinging palm of your hand rather than that of your body. As you walk the aisle, your fingertips seem to move periodically toward LA, then toward New York; that is, along with then opposite to the direction of motion of the plane with respect to the ground.

The ancient Greeks dreamed up geometrical models to explain why planets appear to wander and why they make retrograde loops. One of the earliest accounts had each planet pinned to its own rotating sphere, the axis of which was attached at the approximate angle to the most distant sphere, the one that contained the fixed stars. This outer sphere spun about the earth in twenty-four hours, while a planet's sphere turned on its own axis in the reverse direction in a longer period. The combined effect of such a model duplicated the compound motion I have just described. A later model had each planet move on its own orbit, the center of each describing a second orbit about the earth. Seeing the motion of the planet from the surface of the earth then becomes rather like watching a piece of chewing gum stuck to the wheel of a bicycle passing by. As the gum passes over the top of the rotating wheel, it lurches forward in the direction of motion of the bicycle, but when it moves around close to the pavement it appears briefly to go retrograde. One of the revolutionary developments of this sort of geometrical modeling consisted of mentally moving the earth, together with all the planets, onto orbits centered on the sun. Now the retrograde motion can be explained by the earth passing a slower planet in orbit around the sun, just as a truck on a highway appears to go backward relative to the distant scenery as you pass it. However, both these schemes are derived from Western astronomical thought, and we have no business afflicting other civilizations with them.

If you want to follow the planets through their complete cycles, a month, even a year out under the stars will never do. But if you are patient enough to allow yourself and your descendants years of collective experience that can be passed on orally and in written form, it might eventually dawn on you that the five wanderers seem to be of two distinct breeds. On the one hand, free-ranging Mars passes all the way around the nighttime sky in about two years, Jupiter and Saturn in a little more than a year. Sometimes they are seen as morning stars, close to the sun in the predawn sky, and at other times they are evening stars, setting in the west after the sun. Often they ride high in the sky at midnight. At other times they disappear from view, lost in the glare of the sun. On the other hand, in stark contrast, Mercury and Venus always stay close to the sun; Mercury never strays more than two handspans, Venus no more than three from the sun. Each bobs like a yo-yo, back and forth, toward and away from the sun, Mercury completing a cycle about three times a year, Venus in a little over a year and a half. The effect is especially noticeable for Venus, not just because it is exceptionally bright but also because Mercury, its imitator, spends much more time lost in the glare of the sun.

And this is what makes the great white light of Venus so special—worth following more closely—because we can discover through it the rich detail that our ancient and far more patient counterparts perceived and wove into their mythologies. But to do so we must free ourselves from the sun-centered model of the universe, which insists upon reducing the observed motion of Venus in the sky to an imaginary orbit about a sun that slowly glides along the horizon.

You are standing under a clear sky looking east in the midst of morning twilight. As time passes and the sky grows brighter, stars lift out of the horizon and glide sideways off to the south. Just as the last starlight fades from view in the reddening sky, you glimpse a bright light clearing the horizon. In a moment , however, it is gone, swallowed up by the impending daylight—Venus's first appearance as morning star. Next morning you return at the same hour. This time Venus rises a few moments earlier and moves a bit higher in the sky before it vanishes and the sun takes its place. Next day it is still brighter, yet higher.

A typical track showing how Venus moves in the morning twilight over several months is shown in Figure 2-4. The symbols depict where it was last sighted in the vanishing morning twilight at two-week intervals. Notice how every day during this period Venus seems faithfully to announce the sun's arrival; it begins its morning aspect by dramatically bursting upon the celestial scene—that first appearance out of the solar glare astronomers call heliacal rise. From morning to morning it gradually stretches away from the sun until it reaches its maximum extension, when it stands at its highest point in the predawn sky. Then it begins to snap back toward the sun and is visible in morning twilight ever more briefly, until it makes its last morning appearance, disappearing once again in the solar glare. Gone from the predawn sky for several weeks—an average of fifty days—it repeats the process in the evening sky, hovering like a guard dog in the encroaching darkness over the spot where the sun went down. It acts as an evening star, going from first appearance to last in about 263 days. This time it is out of sight for only several days before returning to the morning sky to repeat the whole process.

Historically, Venus's disappearance periods have been as significant as its appearance intervals. Cultures on opposite sides of the globe reckoned, even canonized the Venusian disappearance, particularly the one preceding the first appearance in the east in the morning sky. All along its track Venus brightens and fades, reaching maximum brilliance only days after it first arrives on the scene.

But all planets do not behave alike. Suppose, for example, that you watch slow-moving Saturn and begin by sighting its first appearance in

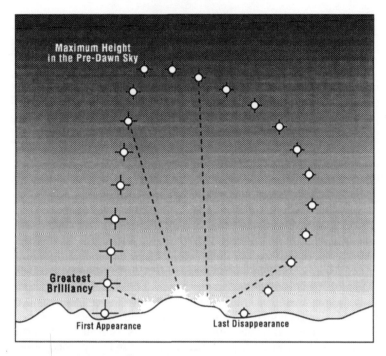

FIGURE 2-4. THE QUINTESSENTIAL PATHS OF VENUS
Standing in the same spot and following the motion of Venus from night to night at the same time of night, this is what you would see. Symbols spaced at two-week intervals show where "Morning Star" Venus would just disappear at the end of twilight in the east over several months. As Venus makes a sweeping arc across the twilight sky from first appearance at horizon through maximum altitude to last disappearance below the horizon, the sun, connected to it by a dashed line, creeps slowly along the horizon. The sizes of the spikes on the Venus symbol denote relative brightness toward its winter standstill. This looped track is one of only five distinct shapes the course of Venus can take in the morning or evening sky. (Drawing by Ellen Peletz)

the predawn sky. If you repeat this observation from morning to morning, you would see the planet separate farther and farther from the sun, rising earlier and earlier each day. But, unlike Venus, Saturn would not spring back and return to the sun. Instead you would see it creep from constellation to constellation all the way across the heavens. Eventually, a few months later, it would stand high in the south at sunrise and, in a few more months, approach the western horizon opposite the light of dawn. About then it would execute its little retrograde loop against the stellar backdrop. Finally it would disappear in the west at dawn, only to return to the evening sky, where you would be sure to see it popping up in the east just after the sun goes down in the west. Its first appearance as evening star would be followed by a full course across the night sky and conclude with its predictable plunge back into the setting sun. All of this would happen in a little over a year.

Mars and Jupiter would mimic Saturn but on schedules of their own. In a single cycle each of these three planets makes a grand entry, a sweep all the way across the nighttime sky, and one disappearance interval in the sun's morning or evening glare. Not so for Venus, which has two equal periods of appearance punctuated by two unequal intervals of disappearance—time spans that alternate with each other endlessly: on for about 263 days, off for 8 days, on for another 263 days, off for about 50 days, and so on, over a total averaging 584 days.

This 584-day-cycle—call it the Venus year—meshes perfectly with the length of the year of our seasons—365 days—in the ratio of 5 to 8, like two gears of a celestial timepiece. To the careful eye, this means that any visible aspect of Venus timed relative to the position of the sun will be repeated almost exactly after eight years. For example, if Venus first appears as morning star on the first day of autumn 1992, it will do the same very close to that date in 2000.

A seasonal index like this could be useful to any practical-minded people who kept time by a solar-based calendar, especially if they were attracted to tidy whole-number ratios. As we shall discover, the ancient Maya of Yucatán were such a culture, and they monopolized and revered the Venus cycle, recognizing that this curious four-phase motion harmonized with a host of other natural periodicities, including the time between the conception and the birth of a child. The eight-year cycle also envelops a whole number of twenty-nine-and-a-half-day lunar months—ninety-nine to be exact. This means that whatever phase of the moon accompanies the first appearance of Venus at the autumn equinox in 1992, that same phase will repeat again about the first day of autumn in the year 2000, along with the return of Venus. This coming together of cyclic periods may seem insignificant to us— it scarcely matters to us what day of the week coincides with New Year's Day—but, as we shall see, for cultures whose system of timekeeping was based on repetitive cycles going all the way back to their mythic creation, such commensurations in the wandering of their celestial deities would have been major discoveries capable of disclosing the secrets of the gods.

We are simply not used to watching the planets this way. We track them among the stars; we look at them through a telescope, but our modern society has no real interest in noting where they enter the sky from the horizon and where they leave us to pass back into the underworld. Later, however, we shall see that people in other times and places fashioned their own cosmoses out of quite different planetary perceptions.

Venus reveals further secrets when we follow it in the same way over a full eight-year cycle of morning and evening star curves like the one in Fig. 2-4. Imagine you were viewing Venus in a series of these time-lapse sequences from a building in an ancient ceremonial center. If you watched Venus over several years, you might come to realize that the serpentine curve, which required several months to execute, can take on various looped shapes. And, if you compare them, you will find only five repeatable forms of the course of Venus. The interval between the execution of a pair of sky paths of similar form seen at the same time of night is eight years or 2920 days, made up of five Venus cycles, each averaging 584 days.

Like the sculpted Aztec calabash I described in Chapter 1, these patterns of recognition have no function for us. But what about the astronomers of other cultures? As I will demonstrate in Chapter 5, these direct visual perceptions, largely obscured in contemporary astronomy, were indeed noticed by many of our Old and New World predecessors, especially the grouping of Venus cycles in fives.

Another directly observable planetary cycle, also unrecognized in modern astronomy, consists of the change in position of daily appearance and disappearance of the planets at the horizon. As we might expect, Venus executes an annual horizontal oscillating motion that more or less follows the sun, and the limits that it reaches along the horizon, though spaced a bit farther apart than the solstices, repeat every eight years. Likewise, the other planets also oscillate like a pendulum and with just about the same amplitude, each in its own period. But because Venus is always near the sun, we might well suspect it was the one watched most closely. I believe this back-and-forth motion goes a long way toward explaining the oft-mentioned references in ancient writings to seasonal phenomena ruled by Venus. In Mayaland for example, the deity is tied to rain prediction and the growth of maize. But how could a farmer possibly associate the behavior of Venus and the welfare of his crops? Rainy seasons occur on a yearly basis, but the 584-day Venus cycle is more than a year in length. We have a choice: Either the connections they wrote about had no real physical basis, or the ancients saw Venus behave in a way that has escaped our detection.

The fault lies in ourselves and not in the stars. Perspicacious observers who regularly get a good look at the local skyline will discover to their surprise that the place of disappearance (or appearance) of Venus can be easily correlated with the length of the disappearance period, and this in turn can be linked to the season of the year. In the Mayan tropics, for example, when Venus disappears during the month of August, it can remain absent from view for more than twenty days, whereas in February it is likely to be gone for a few days until its morning return. In rare circumstances it is even possible to view Venus as evening and morning star on the same day. But if we average out disappearance intervals of Venus over several years, the result is about eight days, precisely the period assigned to that particular aspect of the planet in the ancient Mayan calendar.

Ancient astronomy was about trying to know the future—being able to predict well in advance where the powerful celestial forces and harbingers of things to come would be positioned well in advance—so people could get ready. This cycle of Venusian horizon appearances likely provided clever sky watchers with a precise device that had implications reaching well beyond setting the time for planting or anticipating rain. For if the time of appearance and position of Venus at the horizon vary in a regular way with the season of the year, people with such knowledge could predict exactly how long it would be before Venus would return once it had vanished, as well as where it could be seen when it did return. Knowledge is power, and, as you will see, the corpus of ancient inscriptions, iconography, and architecture supports an abiding concern for such important detailed matters.

The brightness of the planets also fluctuates. Venus is the brightest of all; there are even instances when a careful observer can see it in broad daylight. A crowd of peasants who had assembled in Luxembourg to hear a speech by Napoleon is said to have become attracted to the image of Venus in the noonday sky glimpsed off the corner of the palace from the balcony onto which the dictator was about to emerge. Clever Bonaparte seized the day by proclaiming this apparition of the planet a good omen, foretelling his conquest of Italy.

Mars exhibits the greatest variation in brightness, occasionally rivaling Venus, then slipping to the luminescence of a moderately bright star. Astronomy today attributes the variation in brilliance of the planets in large measure to their distance from us. For example, when the red planet Mars is opposite the sun from earth, it is distinctly fainter than it appears when we are both on the same side. There have been some close passages of earth by Mars, such as that of 1956, when the Martian red glow becomes unusually intense. As you will learn in Chapter 6, when Venus is nearest the earth, its disk, visible through a telescope, is six times wider than when it is farthest away. Seen just a few weeks before last disappearance from the evening sky in the west or just after first appearance in the morning sky in the east (as marked in Figure 2-4), Venus attains its greatest brilliance. Curiously, this also marks the time when Venus, viewed through the telescope, is in the crescent phase. Even though most of the disk is in shadow at this time, the area of the illuminated sicklelike portion is greater than the fully illuminated disk that can be seen when Venus lies on the other side of the sun. The sheer size and brightness of this thin disk raise the possibility that Venus could have been viewed as a crescent without the aid of a telescope. (In Chapter 5, I will mention a few such sightings, old and new, that have been reported.)

Occasionally, the five wandering deities could be seen passing one another in the sky, closely gathering about significant stations in the zodiac—alongside the Pleiades, in the mouth of Scorpio, or on the horns of Taurus the Bull. These planetary conjunctions were duly recorded in written documents so that future encounters could be predicted and so that the hands of time could be set back to the midnight of creation day.

In this chapter I have briefly sketched out how the planets behave as seen from your backyard, and I have concentrated on those phenomena that will come up in discussing the cultural appreciation of the planets in later chapters. If the motions I have described in this little naked-eye catalog seem foreign and exotic, perhaps it is only because to see them we need to disregard the knowledge we have acquired from modern textbooks about how we ought to conceive of celestial motion—in terms of orbits centered on a place 93 million miles from where we live. In this book I am concerned instead with what can be witnessed by everyone, regardless of his or her geometrical preparation or mathematical understanding. The only requirement is a careful and patient eye.

You have seen, for example, that the real Venus—the one visible in the sky—moves not on a circular nor even an elliptical orbit, but instead on a sinuous curve as it guards the sun closely, twisting and turning about it, alternately darting toward or fleeing from it—never very far ahead or behind it. Whether it be followed by marking the place on the horizon where it appears or disappears or by counting days of disappearance and appearance in its fourfold cycle, the Venus rhythm beats in perfect correspondence with the seasonal solar cycle in ⅝ time.

Looking at the dizzying motion of the celestial figure skating in Figure 2-4, you can better appreciate the difficulties encountered by astronomers in the Western tradition, who sought to transform what they saw into maps of circular orbits, first in an earth-centered and later in a sun-centered framework. And you can appreciate the eureka of Kepler or the triumphant elation expressed by Ptolemy in his *Syntaxis Mathematicus* at the mere possibility of achieving a solution: "In studying the convoluted orbits of the stars my feet do not touch the earth, and, seated at the table of Zeus himself, I am nurtured with celestial ambrosia."

Before I open the gate to ancient celestial storytelling, we must remember above all that our particular problems and questions about the sky may not have been shared—may not even have been a matter of passing interest—to many of the people on whose planetary conversation you are about to eavesdrop. Even though our modern minds have long since taken flight beyond the horizon to the abstract reference frame of star-fixed extraterrestrial space, our eyes can still look backward in time. We need only to turn off the lights and go outdoors to discover what our predecessors saw in the sky.

CHAPTER 3

MYTHOLOGY:

NAMING THE IMAGES

People had insight in the past about a form of intelligence that had organized the universe and they personalized it and called it "God."
—PHYSICIST DAVID BOHM, QUOTED IN AN INTERVIEW
WITH RENÉE WEBER, P. 21

A CONTEXT FOR MYTH

Julius Caesar was very attached to Venus. He often spoke of his divine descent from her. He wore her figure on his ring, ordered her effigy printed on coins, even built and dedicated a temple to her as part of a Venus cult he initiated. She was on the one hand his *Venus Victrix*, the mother of victory in battle, Caesar's watchword when he went to war, and on the other *Venus Genetrix*, mother of all men. "Mother of Aeneas and his race, delight of men and gods, life-giving Venus, it is your doing that under the wheeling constellations of the sky all nature teems with life" read the opening lines of Roman author Lucretius's poem *On the Nature of the Universe*. Aligning the warrior tradition with the creative powers of fertility was indeed a wise choice for the emperor Caesar—but why Venus? What imagery validates placing the power to wage war alongside creativity? How can both be contained in the same goddess? The answer lies in Rome's ancient Middle Eastern past, and it is a safe bet Caesar never even knew the connection.

I have assembled in Figure 3-1 a set of representative planetary images taken from myths—traditional stories about nature personified—from all over the world. Some of the images and their names can be connected directly to the physical appearances of the wandering lights in the sky discussed in the previous chapter. Others leave us wondering about origins. This chapter's goal will be to convey how some of the metaphors

**FIGURE 3-1. IMAGES OF PLANETARY DEITIES
FROM AROUND THE WORLD**

3-1a. In ancient Babylon, Venus was part of a triad with sun and moon (Portion of the diorite *kudurru* [a kind of stela] of Melishipak. Louvre 64Y33).

3-1b. Ancient Middle Eastern winged Venus, second humanoid figure from left, announces the rising sun god, whose head just begins to pop out of the underworld. The water god, at the right, comes to her aid (Cylinder seal of Adda. British Museum, B.M. 89115).

3-1c. In Persia, Venus is Anahita, the very image of feminine seduction exemplified in Ishtar's evening star aspect (Hermitage Museum, St. Petersburg).

3-1d. Mars was Nergal, the powerful and warlike god of the underworld in ancient Babylon (Drawing by Faucher-Gudin, Maspero, p. 691).

3-1e. Nebo or Nabu, The Wise, was the Babylonian version of Mercury. (Drawing by Faucher-Gudin)

3-1f

3-1g.

3-1f-h Medieval transformations of pagan celestial gods exhibit Christian overtones: **(f)** Saturn, Jupiter, Mars, and Venus appear together in a medieval manuscript (Vienna, National Bibliothek ms. 2378, fol. 12v Michael Scot); **(g)** Wise Mercury becomes a scribe (British Mus. Add. ms. 16578, fol. 52v); **(h)** Just Jupiter is changed into a monk (Florence, Campanile di Santa Maria del Fiore).

3-1h.

3-1i. Quetzalcoatl (Venus) supports a Mesoamerican sky, that bears planetary symbols (Vienna Codex. Graz: Akad. Druck-Ü Verlag, Codices Selecti v. 15, p 47).

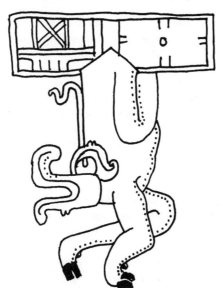

3-1j. Mars, a long-nosed beast, hangs from a sky-serpent band (Dresden Codex). (Drawing by Chelle Wolboldt.)

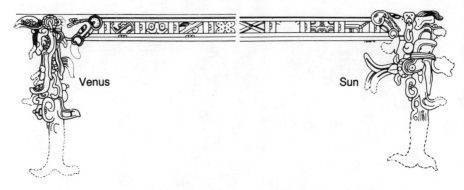

Venus

Sun

3-1k. One of many Mayan double-headed sky serpents carved in stucco and dedicated to the worship of the sun and Venus. Note the sun sign at one end and the Venus signs at the other on this version from the ruins of Palenque (Schele and Miller).

that lie behind the names derive from the discovery of that hidden likeness between what people saw a planet do and the aspects of their lives they sought to express. How did visible aspects of the planets get translated into mythology? Answering this is a complex task, for the names are many and varied and, as you will see, they can be drastically altered by the contact of one civilization with another.

Designating the planets by name and imbuing them with omens lay at the foundation of the astrally based religions practiced by nearly all our cultural predecessors. As above, so below. For them, what happened in real life was mirrored by what took place in the sky. The drama overhead constituted a parallel plane of existence on which people here on earth could reflect and examine human behavior. They personalized nature, as physicist David Bohm puts it in the chapter epigraph. The conquest and absorption of one city by another could be foretold and retold by the perceived interaction among those celestial deities to whom the respective cities paid homage. Omens—words from the mouths of heaven—were prescriptions that served as motives and potentials for human action. If things did not work out, there would be other predictions. Even if the sky were different in different places, people were still the same.

When we try to dismantle an omen in search of its underlying causes, we often can lose sight of the meaning it was intended to convey to the true believer who simply experienced its sign. For those intended to receive them, omens were more like miracles. "Miracle is simply what happens; insofar as it meets people who are capable of receiving it, or prepared to receive it, as a miracle," says theologian Martin Buber. Historians have attempted to reconstruct the real events that gave rise, for example, to the story of Moses's parting of the Red Sea. They focus on the question of what natural process, what combination of wind and water could have created an unusually low tide in a shallow bay at just the right time to permit the Israelites to escape pharaoh's pursuit. More important in human history is not how tides work in the Gulf of Aqaba but rather how the children of Israel interpreted whatever happened. What they experienced was an act of God; it astonished them, and they endured and preserved it.

For those who followed them, that experience became one of the abiding pillars in the edifice of their coming into being as a people. Like miracles, omens are neither supernatural nor superhistorical to those who believe in them. They have objective foundations that can be tied rather precisely to natural events in real time—even if that is not what matters most to us about them.

We need to stretch our imaginations to understand their world. When we do we discover an embedded sensory base for much of the story line that makes up ancient mythology. A simple mythic statement invoked through name and title often links directly with the perceivable world. The planetary representations I will discuss were manifestations of very real, often predictable forces of nature. For example, to an ancient Sumerian, life did spring from the earth—the silt washed down from the mountains by the Tigris and Euphrates formed an ever-growing fertile delta in the Persian Gulf where once there had been but a swamp. And the sun really *was* hostile to life, for every year when it blazed excessively, it dried up the newly planted crops and their nurturing waters, bringing famine and pestilence.

Every subtle bend, kink, and turn, every closely watched disappearance and reappearance in a planet's cycle was carefully written into a script of life as rich and complex as an Ibsen play or a Ptolemaic ephemeris, as vibrant and colorful as Homer's *Odyssey* or a Verdi opera. It is our loss if we choose to disregard celestial mythology as purely nonscientific superstition. It has its real elements, and they are well worth reclaiming.

In a few rare instances, where enough documentation survives, it will become possible to examine some of the intricate plots and subplots in these celestial stories. When we probe, we discover detailed, precise knowledge that Middle Eastern as well as ancient native American people had acquired, through careful sky observation with a scant minimum of technology, about the exact celestial whereabouts and planned routes of their wandering deities.

43

THE POWER OF WORDS

"**S**peech is the most beautiful kind of theoretical magic" reads an old Arab proverb. In his dialogue with God, chronicled in his *Confessions*, Augustine, the fourth-century A.D. archbishop of Hippo in North Africa (now Algeria), asks: "By what means did you make heaven and earth? What tool did you use for this vast work?" After consulting manifold possibilities about how to mold a material universe, an answer finally dawns on him, but only when he has carefully read the first few chapters of Genesis: "And God said, let there be light, and there was light." Concludes Augustine: "It must therefore be that *you spoke and they were made*. In your Word alone you created them."

Still puzzled, and ever the probing inquirer, he goes on: "But how can a word create?" How did God's spoken word result in the act of creation? If words come forth in time, don't they die away? Who heard God's words? Was there a mind, an intelligence, that could initially respond to them? Exactly what words, what syllables, what vibrations, triggered the creation? What was the process? Struggling to comprehend the new religion of Christianity, Augustine was asking questions few had raised before him about how to commune with this abstract, all-pervading God of humankind. And he asked them with the passion of a sinner turned religious convert—which he was. God's words, Augustine concludes, cannot be subjected to this sort of physical analysis. They transcend the laws of time—they are eternal. God's word is uttered not a participle at a time but rather all at the same time— eternally. It is not subject to change.

Today no one would question the power of invisible gravity that holds the universe together. It is part of our stock of common sense—an ingrained explanation shared by all. Yet we have difficulty understanding Augustine's explanation of the power of the word, the divine command. To divine means "to foretell," by word or by prophecy, to enable to happen what happens by divulging *through the word*. Spoken names, words, and numbers, the spoken epithet—all radiated force and emanated energy by themselves. Abracadabra!

In this chapter I want to explore the connections between celestial myth and planetary image through spoken words and to trace the sources of some of the appellations once given the planets in stories that people told about them. I will show what these utterances evoked in the complex and subtle behavior of the planets against the backdrop of the stars that I mapped out in Chapter 2. But first I need to give you a feeling of what it might have been like to experience the potency that lay behind these names. Like Augustine, we must prepare ourselves for the possibility that we cannot apply our own rational analogies to decode or demystify what we learn.

Since the advent of printing and the expression of masses of information in material form—papers, letters, books, and now computers—direct, face-to-face human discourse has been greatly diminished. Our encounters are more often indirect, mediated by technological devices. We speak not to a human face but through a plastic armature. We interact by fax, E-mail, and answering machine over wire and satellite, receiving clipped messages to which we must later provide thoughtful yet succinct responses. True, we issue verbal commands to our children, which, when obeyed, almost surprise us: "Be home by eleven"; "Don't put your hand on that hot stove." We hope that our prayers, which we rarely utter aloud, will be heard and answered, but it would be uncommon to expect direct divine intervention. "Your wish is my command," says the genie, but our common sense is devoid of genies. It is unimaginably different from that in which some of the peculiar planetary names I will talk about were conjured up. How can they make sense to us?

Numbers once had power too. In one of his seventeenth-century *Dialogues*, Galileo denounces the ancient Greek notion that number can determine, just by itself, how matter behaves. After a lengthy argument by his aptly named detractor, Simplicio, who believes that the number 3 is perfect because all complete and whole things in the world have three dimensions as well as three parts—a beginning, a middle, and an end—Galileo replies: "I feel no compulsion to grant that the number three . . . has a faculty of conferring perfection upon its possessors."

Galileo's straw-man opponent believed in the number theory of the Pythagorean philosophers of fifth-century B.C. Athens, who insisted that, like word and name, number—not matter and space—was the ultimate reality in the universe. What compelled them that failed to convince Galileo? For one thing, their arithmetical basis of musical harmony recognized that consonant chords can resonate from plucked strings only if their lengths stand in the ratio of small whole numbers, such as 2:3 or 4:5. It was the mathematics, not the strings, that created the harmony. The elegance of such a relationship was powerful enough to reach across two millennia and influence Kepler. All that remains of this archaic way of thinking are notions such as lucky 7, unlucky 13, and the adage that all good things come in threes. A Greek philosopher once invoked the existence of a hypothetical tenth planet in the solar system just because the nine entities known at the time (seven planets, the sun, and the moon) were but one short of perfect. Today the number 10, though still the base of our mathematical system of numeration, has been stripped of its divine properties.

Pure number also had potency in ancient Mayan thought. Each number was conceived as a god, with particular characteristics: youth or old age, gender, degree of sexual prowess—just about every range of personality trait that could be cataloged. For example, thick lipped, his face spotted with tattoos, the depicted god of the number 2 symbolized death and sacrifice; the wrinkled countenance of number 5 reminds us of the wisdom of old age. Zeroes are represented by full figures with their hands clasped against their jaws. The Maya carved their numbers on tall, rectangular stones called *stelae*. People once stood in front of them and chanted the names of the number gods in the hope that their intervention in daily affairs would lead to a better life. In Mayan society, divine number made the passage of time possible, for the number gods carried the burden of the days, parceled out into units (like our days, months, and years), upon their backs.

Picture books from central Mexico, written before the Spanish Conquest, depict people wearing spoken number-names. These are their birthdays, and they are shown floating astride the figures they represent, each attached to its bearer by a tiny string. Like number, sometimes name alone or a part of the costume to which it is connected may be taken to represent an individual's characteristics, just as the cap I wear may tell where I work or which sports team I favor. But imagine a society in which all of my attributes were foretold simply by giving my name. Modern parents-to-be still consult popular lists of birth names that give descriptions or characteristics supposed to go with each name: Michael and Jessica are godlike; Paul is small; Linda, beautiful. Some are rather obscure: Philip, a lover of horses; Deborah, a keeper of bees.

We do these things, if at all, for amusement. What's in a name? we say. Does anyone seriously believe a spell or an omen can be cast simply by saying a name? How can the title of a person or thing carry direct power or influence—like gravity? This was precisely Augustine's question in his effort to grasp the relation between word and action.

NAME AND PLANETARY ATTRIBUTE

What escapes us is that Mayan, Egyptian, and Babylonian alike believed that they lived in an animated universe, a breathing, teeming, vibrant, and interactive surrounding. The gods of the ancient Near East, for example, were not personages who guided nature's forces or programmed its laws. These deities began as the actual attributes or properties of the material elements to which they gave their names. Esharra, an earth god, was the manifest fruitfulness of the land that made for a bountiful harvest and fat cattle. Sky god Nergal (Mars) was the red feverishness of the summer sun that destroyed crops; Merodach the youthfulness of the spring equinox sun; Dumuzi the sun at the beginning of summer; and Ishtar (Venus) the returning greenness of the grass after winter's frost and summer's scorching heat. These attribution deities were interdependent. They did not simply work together automatically, like cogs in a vast, indifferent machine.

People had an active role to play, too. They talked to the stars, listened to the planets. They commanded and evoked, restrained and constrained, made incantations, pressed their ears to the oracle. They saw themselves as mediators in a great universal discourse. At stake was the battle between fate and free will, between body and soul. When you spoke the name of the wandering lights in the sky, you did not do so frivolously. Your need to summon all your senses was as vital as that of any modern naturalist. And the power of mind we call imagination required to seek hidden relationships between the characteristics of the gods and particular attributes of the natural world—this was every bit as cultivated among the ancients.

Our predecessors employed the same discovery principle that Newton used when in a flash he unified the pale moon and that falling apple. But, unlike Newton, they were motivated by a desire to know how to mediate an alliance between inherent power and direct physical appearance, between knowledge and action. Today we may attribute a planet's color change to an atmospheric effect, a shift in position to a dynamic effect, an alteration in brightness as a distance effect. But the ancients carefully watched the color, brightness, position, and movement of the planets because they believed all of these considered together were indexes of that power they sought to influence through a dialogue. By the process of discovery, celestial bodies took on divine qualities.

So much has been written about the connection between the moon and the feminine principle, of the association between the lunar phase cycle and the menstrual period. The name of the sun conjures up many images, too: the strong will and dominion that his brilliance has exerted over the world, his goodness and justice, the ferocity of his overhead rays in midsummer, the unpredictable maleficence that came when his face was suddenly darkened in eclipse. A Spanish chronicler of the Aztecs tells us:

> *When this came to pass he [the sun] turned red; he became restless and troubled; he faltered and became yellow. Then there were a tumult and disorder. All were disquieted, unnerved, frightened. Then there was weeping. The common folk raised a cup, lifting their voices, making a great din, calling out, shrieking. There was shouting everywhere. People of light*

complexion were slain (as sacrifices); captives were killed. All offered their blood; they drew straws through the lobes of their ears, which had been pierced. And in all the temples there were war cries. It was thus said: "If the eclipse of the sun is complete, it will be dark forever! The demons of darkness will come down; they will eat men!"[1]

But all sky omens need not arise from sudden change. If we think about it, there are good reasons for translating normal solar behavior into a concept of justice, for is justice not based on constancy and consistency, on day-to-day reliability? Even if on occasion the sun became dimmer or more ruddy in appearance, of all the objects in the sky, he above all was the most constant and dependable.

The moon is somewhat more changeable; certainly she puts on different faces, and she wanders and wobbles a bit more than the sun. Isn't this exactly the way we might expect a male to characterize a female? (We must not forget that, regardless of how we feel about it, when we look at the old texts we are reading history from a decidedly male point of view.) All the moon's phenomena had meaning: her horns, her royal cap—the so-called earthshine or semi-illuminated dark portion—the halo that sometimes encircled her. When the ring around her was broken, it meant escape. The Babylonians said that the sheep then got out of the fold.

Venus was given a host of names. She was called Ishtar in Chaldea, Nabu in Babylonia, Anahita by the Persians, Benu by the Sumerians, Astarte and then Aphrodite by the Greeks—all feminine appellations. The Greeks also recognized Venus's dual aspect, referring to it as Phosphoros in the morning and Hesperos in the evening, even though they did not distinguish two separate celestial bodies. Later the Romans named these aspects Lucifer and Vesper. In ancient Mesoamerica, Venus was a male, Quetzalcoatl (feathered serpent); to the Maya he was Kukulcan. Our Hawaiian ancestors named the planet Hoku-loa, the Tahitians Ta'urua.

Wasp Star, Red Star, Great Star, Lone Star, Lord of the Dawn, Home of the Love Goddess, Proclaimer, Companion to the Royal Inebriate, Bringer of Light, Satan himself—all these titles were given this single source of light by imaginative people from various epochs and corners of the world. But what did these names mean? Where do they come from? Bringer of Light and Lord of the Dawn are understandable enough, for I have already charted how often Venus does precede the rising sun. But why has Venus variously been linked with the highest ideal of feminine beauty, love, sexuality, death, resurrection, deceit, and war? What tangible properties of the planet Venus might have provoked the stories in which she received star billing?

At least part of the worldwide Venus nomenclature arises out of a natural curiosity about what the afterlife is like, a subject I will explore in detail in the next section of this chapter. Where do the sun, moon, and stars go when they pass over the western horizon? What journey do they make between their western disappearance and their reappearance in the east? The unseen world, often thought to be the place of departed souls, the occult side of human nature forever hidden from our eyes, played a major role in ancient cosmologies.

"What happens in the underworld?" wonders an ancient Egyptian priest in a pyramid text of the Old Dynasty. "How may we acquire knowledge of what the souls do, of how Re [the Sun] is glorified at night, of what Re says to the souls when he meets them every night? Over what path does he pass?"[2] Because Venus, the Lone Star, enters the western gate of the underworld so vividly and so frequently (five times in eight years, as we learned from our discussion in Chapter 2), this brilliant and far-traveling luminary who brings food to nurture the sun in the underworld is the logical one to ask. It is he who will bring back messages of what takes place there, and he knows exactly how it will affect us. The Lone Star is the carrier of omens from the netherworld.

The Maya called their underworld Xibalba. There lived the Lords of Death, the ones who sought to bring misery upon the human race. Deaths to be feared most were those borne by disease or other afflictions. Maybe that is why the underworld gods were given weird-sounding names, such as Stab Master, Pus Master, Jaundice Master, and Blood Gatherer.

The Mayan culture of the Yucatán peninsula rose to prominence around the beginning of the Christian era. We still know relatively little about these people because the New World archeological record has only begun to be unraveled. While Heinrich Schliemann dug Troy and Sir Arthur Evans unearthed the Mycenaean remains of Crete, the Mayan ruins lay undisturbed, blanketed by the Guatemalan rain forest. In the twentieth century, however, we found firmly embedded in the topsoil of America's pre-Columbian civilization the same sort of celestial worship—almost to a tee—that we had encountered in ancient Babylonia: monumental religious architecture, mythological creation episodes featuring the sun, moon, and planets in starring roles. At the height of Mayan urbanism, separate city-states held local power. We have only recently deciphered the colorful names of their rulers from the phonetically based hieroglyphs written on their funerary monuments: Stormy Sky of Tikal, Bird Jaguar of Yaxchilán, and New-Sun-at-Horizon of Copán.

In the *Popol Vuh*, the creation story of the old Quiché tribe of the Maya, Venus is a twin who goes into the underworld with his brother the sun to battle these lords of pestilence. In one curious episode, Venus defeats the evil lords by tricking them into offering themselves for sacrifice. He demonstrates his power by sacrificing his twin brother, tearing out his heart, and then, by voice command, bringing him back

49

to life. So ecstatic are the lords with such legerdemain that they plead to Venus—"Do it to us!" Venus indulges them, but, cleverly, he only completes the first half of the process. Were it not for Venus's cunning actions in the netherworld before the dawn of history, the Maya say, the world would be far worse off with disease than it is today.

In the story the sky deity makes five journeys to Xibalba, one to match each of the five unique paths that Venus takes in the twilight sky. The purpose of the journey is to sow the seeds of creation—to make the world ready for the first dawn. Every time Venus appears as morning star to announce the arrival of his twin brother, the sun, he resketches for us the cycle of creation that led to the present-day Maya lineage. By offering blood to the morning star, the ancestors of today's only partially Christianized Mayan people paid back the sanguinary debt they owed one of the deities responsible for their presence here on earth.

Ancient Mayan painters and sculptors placed symbols of the sun and Venus on opposite ends of a two-headed serpent, Caan, that represented the sky. They draped the Caan serpent in stucco over the doorways of religious temples (Figure 3-1k) and delicately painted it on ceramic plates. In one of the surviving picture books from Central Mexico, a Venus deity (Figure 3-1i;) supports a segment of the serpent sky that is studded with the symbol of his bright star. The sinuous image of the two-headed sky serpent offers a graphic depiction of the way the imaginary line connecting Venus above the horizon with the sun below can be followed through time. When Venus, the front end of the cosmic monster, first appears in the morning, he reenters our world from the underworld realm of Xibalba. Having been there, he always brings with him an attitude or feeling to impart to us, call it an omen— a sign from his mouth. And if we pay close attention, we can learn what that is. The first morning vision of Venus graphed out in Figure 2-4 is to the believer a celestial annunciation, a word of mouth based on the special connection between Venus and the sun.

In the ancient Middle East, where Venus was called Ishtar, there is a long history of worship of the goddess who evoked the power of the dawn. In Sumeria, or southern Mesopotamia—today's Iraq—about the middle of the third millennium B.C., political power began to be concentrated in large population centers. The archeological record attests to increased improvements in farming technology, which in turn had led to overexploitation of the Euphrates Valley. It all happened at a most inopportune time, because a period of extended drought had just begun to take hold across the Middle East. Out of the whims of nature and human ingenuity, the city-states of ancient Sumeria were born. Those that thrived were the ones built and ruled by the most clever leaders—those who knew how to fortify and defend their territories, who were the most aggressive campaigners in the never-ending quest for ever-shrinking arable land. One could have predicted that the town chiefs would erect monumental places of worship to the gods of the natural forces who had treated them so harshly but who held the future in their hands.

Uruk (Erech, from the Bible) was one such successful city. Today it is the Iraqi city of Warka, about halfway between Baghdad and Basra. It was ruled by the sky god Anu and the love goddess Ishtar or Nin-si-anna (the Lady of the Defenses of Heaven) in Semitic Sumeria or Inanna (Queen of Heaven) to non-Semites. Each deity possessed his or her own ziggurat, a huge temple complex consisting of football-field-length courtyards, paired sanctuaries, square towers, private rooms, monumental platforms, and altars—a place big enough to house a whole retinue of specialized cultic priests, overseers, and scribes.

The first syllable of Ishtar's name is probably derived from the Sanskrit *ush*, meaning "a burning or fire." *Ush* also came to mean "east," the direction to which worshipers turned their faces in order to feel the rays of the bright sun god, both powerful and nurturing. Later, east became the cardinal axis about which most early Old World maps were constructed. It marked where the sun rose on the equinoxes, the first days of autumn and spring. Our word *orientation* means "easting," and most old European cathedrals face that direction. The pre-Christian worshipers upon whose pagan temples these edifices now stand needed to be sure that every time they faced the cosmic axis and uttered Ishtar's name they would soon feel within their breasts the power of dawn, of fire, of creation and fertility.

When the Sumerians spoke to Ishtar they sought to draw out her feminine sensuousness:

> *Ishtar is clothed with pleasure and love*
> *She is laden with vitality, charm and voluptuousness.*
> *In lips she is sweet; life is in her mouth.*
> *At her appearance rejoicing becomes full.*
> *She is glorious; veils are thrown over her head.*
> *Her figure is beautiful; her eyes are brilliant.*[3]

Those who called after Venus-Ishtar as she disappeared into the underworld were not primarily concerned—the way we are—with seeking out some abstract set of predictive laws that governed her physical behavior. Their principal activity was to discover how perceivable planetary changes might be used as part of a stage set to act out human concerns. If the modern astronomers' sky quest has developed into a search for cause behind effect, that of the ancient Sumerians was a game of seeking correspondences and associations. They asked different questions in their dialogue with nature: What changed Ishtar? Did her affair with the underworld gods cause her to exchange her portents of fertility for those of famine? And beneath it all, the bigger question: What is this thing called love? "Venus teases lovers with images," says Lucretius, but we from the distant present wonder, How is love's image duplicated in Venus's appearance in the sky?

The Roman historian Pliny offers a materialistic explanation of Venusian fertility. Being so close to the horizon at both of its first risings, Venus scatters a genital dew that fills the sexual organs of the earth and stimulates those of animals. It is this association that ties the Great White Light to the sexual act and is responsible for nighttime sexual reproduction. Venus near the earth, dew on the grass—both occur at the same time.

Because they once were earth gods, all the planets in the Sumerian pantheon had terrestrial dwelling places. Ishtar's home, as we have seen, was a place that radiated sexuality and fecundity. On the contrary, the dwelling place of Mars (Nergal) was a violent domicile that generated the malevolence associated with the war god. Pliny attributes Mars's fiery redness to its proximity to the sun, which can be deduced by the fact that it moves faster than Jupiter and Saturn. The names Assyrians gave to Mars also suggests anything but beneficence and dependability. He was the pestilential one, hostile and rebellious. War was another of Mars's aspects, and some Assyriologists have suggested that this may have been associated with his blood red color, especially when he lies low over the land. Or is it the erratic motion Mars exhibits, well beyond that of the other planets? (Kepler spent his whole life at war with this planet.)

Saturn was Ninib to the Sumerians, seen as a phlegmatic old man who lumbered ever so slowly across the celestial vault. The Roman historian Diodorus, who wrote about them more than a millennium after they flourished, says the Assyrians called Saturn the Star of the Sun because, of all the wanderers, it spent the longest time plodding through the region opposite the sun. Saturn also received the designation Lu-Bat, the steady one, for he could be counted upon, more than any of the other wanderers, to be present in the night sky just as the sun was in the day. Because he moves thoughtfully, steadily, deliberately, Saturn's character reflects wisdom and intelligence more than speed and vigorous activity. Because his course emerges as the least deviated, pagan societies of the later Roman Empire linked with him the qualities of fatherliness. As I will show in Chapter 5, Saturn achieved the highest order in the potent celestial hierarchy. By moving in the uppermost realm, a fact again proven by the extreme length of time it takes him to cycle around the zodiac, he occupied the biggest sphere and was therefore accorded the most power.

Nearer to the sun than the cold, remote location of Saturn yet farther than fiery Mars lies Jupiter—Greek Zeus. When we think of him, we think of justice. He became a moderator and, consequently, most fit to rule the celestial gods. He alone held the power to create storms, floods, and earthquakes. As Marduk he was elevated to the position of tutelary or protective deity of the city of Babylon. He also rose to the godhead position in the later Babylonian astral religions as a consequence of that city having gained prominence over its rivals.

To fleeting Mercury, ancient people applied terms such as *burner* or *sparkler*, which visually depict the way the planet twinkles at the horizon. Fox and leopard are associated animals that characterize its aspect as a trickster, for Mercury would always foil one who tried to follow him by hiding and disappearing so frequently.

These subtle associations between name and attribute constitute what one historian of astrology has called "a remarkable alliance between human imagination and the world of appearances."[4] But beyond the mere recitation of planetary properties, some of the elaborate themes invoked by planetary behavior are more interesting to explore in detail. Venus provides but one example.

VENUS DESCENDING

The implications of the descent of Venus into the underworld are so integral and lasting a part of ancient Middle Eastern star lore that both image and meaning are worth elaborating. From these the later Greeks and Romans (and we today) have acquired the popular image of Venus as the love goddess.

Just as they conceived of the region of the Tigris and Euphrates divided into city-states, so did the Sumerians segment their sky into three longitudinal areas. Each of these sky sectors was paired with a region or country on earth named after the gods: Anu to Elam, east of the Euphrates; Ea to Amurru to the west, Enlil to Akkad, sandwiched in between. When a wanderer entered a given sky zone, its omen pertained to the land assigned to that zone; for example, Venus first rising in Anu would bring (generally) an omen of abundance to the land east of the river. As we might expect, the message often was coded in the metaphor of sexual fulfillment accompanied by moral overtones. The greater the frequency of the sexual act, the more new fruits of the vine would be brought forth, and the more we indulge in the sexual act, the more children, the fruits of our vine, do we bring into the world. One omen reads:

> *When Venus stands high, pleasure of copulation. When Venus stands in her place, upraising of the hostile forces, "fullness" of the women shall there be in the land.*[5]

But how did our Old World ancestors draw out of a single Venus image the human qualities of seductive creativity on the one hand and hostile warfare on the other that we read in their surviving documents—qualities they evoked when they uttered the name of the celestial Ishtar? We must look beyond the astronomical tables and calculations with which they soon became preoccupied in precise pursuit of her. We need to listen to their myths, for it is in these tales told around the hearth to people well familiar with the vivid celestial imagery that we learn about how they conceptualized their living gods who trekked across the sky.

Imagine a group of peasants huddled around a fire on a cold desert night. One sings a ballad to the accompaniment of a stringed instrument while the audience taps out the rhythmic beat:

> From the ["great above"] she set her mind towards the "great below,"
> The goddess, from the "great above," she set her mind
>   towards the "great below,"
> Inanna, from the "great above," she set her mind
>   towards the "great below."
> My lady abandoned heaven, abandoned earth, to the
>   nether world she descended,
> Inanna abandoned heaven, abandoned earth, to the
>   nether world she descended,
> Abandoned lordship, abandoned ladyship, to the
>   nether world she descended.
> In Erech she abandoned Eanna, to the nether world
>   she descended,
> In Badtibira she abandoned Emushkalamma, to the
>   nether world she descended,
> In Zabalam she abandoned Giguna, to the nether world
>   she descended,
> In Adab she abandoned Esharra, to the
>   nether world she descended,
> In Kish she abandoned Hursagkalamma, to the
>   nether world she descended,
> In Agade she abandoned Eulmash, to the
>   nether world she descended.
> [one city deleted][6]

This text, which has been passed down to us on thirteen clay tablets from second-millennium B.C. Sumeria, tells the full story of the descent of Inanna into the underworld. No one knows why she originally abandoned her seven celestial cities, their temples, and the parts of the zodiac they represent. Some say it was to reawaken her lost lover from the dead. Whatever the reason, those who worshiped her knew that were she to stay away too long, all the races of humans and species of animals on earth would become extinct, for, in her absence from the sky, their desire and ability to make love would vanish.

Before departing, she instructs Ninshubur, her messenger, that should she fail to return in three days, he must visit the other gods and ask them to intervene and arrange for her rescue. Taking along the appropriate ritual objects and clad in robe and jewels, she approaches the seven gates of the netherworld, a harsh and hostile place where dust is nourishment and clay is food. She confronts the gatekeepers, who insist that she must shed her garments as she passes from gate to gate. One must enter the underworld totally naked, for as all mortals arrive naked into the world, so must they leave it. Stripped of her clothing, she approaches the underworld gods, who immediately strike her dead and hang her from a stake.

Three days pass, and on the beginning of the fourth day, as his sworn duty, Inanna's faithful messenger hastens on his rounds, inciting a general clamor in all the heavens as he goes. The fireside story continues:

> *After three days and three nights had passed,*
> *Her messenger Ninshubur,*
> *Her messenger of favorable words,*
> *Her carrier of true words,*
> *Fills the heaven with complaints for her,*
> *Cried out for her in the assembly shrine,*
> *Rushed about for her in the house of gods,*
> *Scratched his eyes for her, scratched his mouth for her.*[7]

The god of air (Enlil) and the moon god (Nanna) reject the messenger's plea. But the water god (Enki) consents to help. He devises a plan to restore the goddess of love to the world of the living. He will send two sexless creatures into the netherworld, one to sprinkle the water of life, the other the food of life over Inanna's corpse. Out of the dirt beneath his fingernails, Enki creates the emissaries, who travel below and sprinkle the water and food of life on the dead goddess sixty times. The remedy works; Inanna awakens, frees herself, and wastes no time departing the underworld. She retrieves her garments as she passes in reverse order through the seven gates.

Bits and pieces of the Ishtar myth have been traced all the way back to the oldest of all the world's literary epics—the story of Gilgamesh. Every culture has its legendary superman and, like the Mayan *Popol Vuh* and the Greek *Odyssey*, the Sumerian *Gilgamesh* commemorates the exploits of an ancient hero. Perhaps a real person, more probably he was an exaggerated amalgam of real-life characters who lived late in the fourth millennium B.C. The tablets on which this tale is written were dug out of the ruins of a palace in Nineveh, and although they date to the seventh century B.C., we know from other documented cuneiform writings that various versions of Gilgamesh's wanderings go back to before 3000 B.C., before which the tale probably existed as oral tradition. We know the story today thanks to the later Semitic Babylonians, who adopted an altered version of it as their national epic in much the same way patriotic America acquired and embellished the notion of democracy from the Greek city-states.

Gilgamesh, forerunner of Hercules, lived in the city of Erech. With his sidekick Enkidu, he set out on a series of godly escapades. In one episode he comes to the rescue of Inanna-Ishtar. Once upon a time, as the story goes, the Queen of Heaven found a tree on the banks of the Euphrates that had been uprooted by a flood. She took it back to her palace, where she planted it in her garden, hoping someday to use its wood to make a beautiful throne for herself. But once the tree grew up, she was prevented from cutting it down, for an evil snake had nested with its young in the trunk of the tree and a mischievous bird had done the same high in the treetop.

Hearing the queen lamenting her difficulty, Gilgamesh dons fifty pounds of armor and, wielding his battle axes, comes forth from Erech with his entire army, slaying the snake and causing the bird to flee. The heroes cut down the tree and present it to Ishtar, who uses a portion of it to fashion a drum for Gilgamesh, which she presents to him as a reward. But a series of amorous advances—the evil side of the love goddess—comes with the gift, and Gilgamesh needs to muster all his courage and discipline to resist Ishtar's voluptuous foray.

Later the drum falls through a hole in the earth and becomes lost in the underworld. The ever-loyal Enkidu volunteers to go and retrieve it, even after his master warns him of all the perils—the do's and don'ts such a journey would entail. (You must not speak; you must not use your weapons; you must not clean clothes lest the dead use them to come to life as enemies . . .) In his zeal to please his master, Enkidu disregards the warnings, and he is immediately seized and prevented from returning to earth. Later Ishtar, out of revenge for having been refused by Gilgamesh, is implicated in Enkidu's death, but not before Gilgamesh disguises himself as Ishtar—just the way other planets sometimes imitate morning-evening star Venus—in order to enter the underworld to search for his already-dead companion.

The Ishtar descent myth has its parallel in Greek mythology in the story of young Persephone, daughter of Demeter, the goddess of grain, who is carried off and raped by Hades, god of the underworld. In Roman mythology, Proserpine, daughter of Ceres, is the victim. In any language, the story tells us a lot about people's beliefs concerning death and the afterlife, the power to resuscitate the dead, the relationship between the living and the dead. What lies behind it is a curiosity about the underworld in general—the evils of pillage, plunder, and other misfortune the unknown is capable of inflicting upon the human race and our capacity for retaining or losing what we possess.

Just as the Greeks acquired a knowledge of the sky along with mathematical skills from the early cultures of the Middle East (much of this knowledge having diffused through long and hostile contact with the Persians) they adopted story lines, metaphors, and allegories—a style for relating events that befall us here on earth to escapades of the air, sky, and earth gods, supernatural though uncannily real personae who were the ultimate driving forces behind the natural world both peasant and king confronted. The big questions were all universal: How is the world organized? Where do we fit in? Where did heaven and the underworld come from? What happens when we get there? How do the inhabitants of these unearthly realms behave, and can we learn anything about them while we are alive? The same basic questions, only the names of the players get changed as ideologies are passed on and slowly transformed from culture to culture.

In many instances, our story also reveals tangible visions and events in the natural world that our ancestors embraced as they acted out their myth in oral recital, dance, and song. There were images both storyteller and listener saw that made the fireside story told beneath the stars come to life. We already have acquired a fair idea of why Inanna-Ishtar-Venus is the one who descends and reascends. Venus's dramatic coming and going is an ideal metaphor for death and resurrection. But why is she absent for a specific length of time in the myth? And why is it the water god who comes to her aid?

The modern mind might seem content to sweep such details under the rug of superstition and irrational mysticism. We have become so conditioned to believe that myth can have no basis in observed fact. But by appealing to natural phenomena that actually took place in land and sky in the Fertile Crescent four thousand years ago—by paying attention to qualities peculiar to the planet Venus, for example—we can begin to piece together an empirical side of the mythic coin that complements and enriches the seemingly strange logic of the Inanna-Ishtar myth. If the listener knows what celestial imagery goes with each chapter, the tale truly begins to come to life in the real world, and so we can better share it with our ancient ancestors.

Of the five planets, the first three—Saturn (Ninib), Jupiter (Merodach), and Mars (Nergal)—were generally placed under the parentage of the sun god Shamash, the all-seeing, constant, and dependable guardian of justice. But the other two planets—Mercury (Nebo, Figure 3-1e), and our Ishtar-Inanna-Venus—were said to have descended long ago to earth, where they were paid homage—male Nebo by the people of Borsippa, female Ishtar by those who lived in the city of Dilbat. This pair of sky deities was chosen, as I already indicated, because they alone have always roamed the region of the sky closest to the ground. If not hidden below the earth, each is visible lying low in the west after sunset or hovering over the position of sunrise in the eastern predawn sky. Whereas neither of these ever strays far from the boundary between land and sky, the other three naked-eye planets have always been free to wander all the way across the celestial vault, even the midnight sky.

In all, there were seven moving objects in the sky (the sun, moon, and five visible planets), each on its own spherical shell among the stars. This explains why one would need to pass through seven gates to get to the world beneath the earth. If you sat by the ancient campfire listening to the twilight tale and watching Venus meander across the sky, you could actually have seen the deity becoming paler and paler, shedding her effulgence as she passed ever downward toward her subterranean confrontation. Venus in the story looked just the way Venus does in the sky when it sets.

Why did Inanna specify that after three days and nights she would require assistance, and why did she ultimately get help from the water god, as shown in Figure 3-1b? These two elements of the myth may be tied together by the facts of astronomy and meteorology. Recall that Venus can vanish from the sky for anywhere between a day or two and up to three weeks (the average is eight days). Moreover, in lower northern latitudes on the short end of this spectrum, a three- or four-day disappearance on the average usually happens around February. Now, rainfall in modern Iraq is dependably quite heavy in the mountainous region of the north. There 90 percent of it occurs in the winter months, from December to March. But in the more southerly lowland valleys of the Euphrates and Tigris, where the Sumerians, inventors of the Inanna myth, and later the Babylonians thrived, rainfall is generally lighter, if far more erratic. Destructive floods are frequent, and since at least 3500 B.C. farmers have depended upon irrigation to combat the irregularity and unpredictability of crop production. Rivers are fed strictly by rains that flow from the northern peaks, and they begin to discharge in early March.

The ancient Sumerians must have heeded available celestial warning signs all year round. And, at the flood season of the year, farmers probably paid even more rapt attention to signs in the environment. Did they perceive Inanna's watery connection? Were they aware that a single handful of days, rather than two or three or four weeks of Venus's disappearance was one of the visible signals to anticipate water in the irrigation canals?

If we fail to understand in mythic terms why the water god came to the aid of the goddess of fertility after three days in captivity, the message from the standpoint of the economics of food production rings crystal clear to us. Sumerian common sense saw it both ways. Holding

Inanna in captivity was a way of expressing both the woes of deprivation through famine and starvation and the real images of land and sky phenomena that articulated the farmer's astrally based religion.

It is interesting to note that, comparatively speaking, Venus's habit of dogging the rising and setting sun also had its effect on Chinese celestial imagery. In eighth-century China, for example, white was the color of ghosts, and the brilliance of Venus also mocked the flash of swordly metal. This is probably why the Chinese called the planet the Grand White and the Executioner's Star—a planet that portended deadly plots and cutting edges. When Venus crossed the constellation of the Battle Ax (part of our Gemini), it foretold the clash of weapons; when it entered the Ghost constellation, it was time to execute the vassals. Warriors once stood on the Great Wall following the movements of the Grand White, even at the expense of neglecting to observe the maneuvers of their enemies. When the planet was especially bright—for example, if it could be seen in the daytime—its spotlight aspect indicated an omen of special importance—perhaps the *yin* principle would strongly override the *yang* and a lower-order vassal could rise up against the Emperor of the Sun.

Other planets exacted their own metaphors. For example, the red planet Mars—planet of blood and fire for the T'ang of sixth-century China—was regarded as *fa hsing*, the punishment star. It was especially potent when it passed the namesake constellations of Virgo and Scorpio. Curiously, Antares, the bright red star in Scorpio, also means "rival of Mars" in the Western tradition. Its hot, rosy radiation also warned of drought. An eight-century Chinese omen reads:

> *When the Sparkling Deluder enters the Southern Dipper and its color is like blood, there will be a drought.*[8]

T'ang poetry tells of an apparition of Mars on earth four centuries before, when a young boy appeared among a group of children at play. Bright red rays flew out of his eyes and astonished his newfound playmates, who asked him who he was. "I am the Sparking Deluder," he answered, "and I am here to foretell the victory of the family of Szu-ma over the ruling triple dynasty." With that he shot off like a meteor. In fact, over the next two decades each of the rulers of the triumvirate fell from power.

From tropical ocean to lush valley to mountain village, every subtle bend and turn in the planets' courses brought with it a host of auspices that fueled the great confrontation between nature and humanity. No one questioned whether land and sky were related to spirit and being.

Their questions only explored the nature of that relationship, which, as we shall see in the next section, forged some curious alliances.

DYAD AND TRIAD:
ORGANIZING
THE CELESTIAL FAMILY

L ife has its ups and downs, its love-hate relationships, and we all know that you need to experience pain to appreciate pleasure. Do we define our emotions to reflect extremes that are part of a continuous spectrum, or do we think of them as polar opposites: grief and joy, depression and elation? We develop ways of measuring out all the possibilities for feelings we can ever imagine experiencing—usually on a scale of minus to plus ten.

We structure nature exactly the way we structure ourselves. This is why mythology sometimes portrays planets in pairs (or dyads) and at other times splits them in half—literally regarding them as two separable, opposing yet complementary entities (or moieties). Take, for example, the two aspects of Venus that figure in the Ishtar-Inanna myth. Or our word *hermaphrodite* (Hermes plus Aphrodite), which is a combination of the Greek names for Mercury and Venus. Hardly anyone believes in the Devil anymore, but even those who do find it difficult to think of him as the malevolent half of an otherwise benevolent deity. If, though, you can imagine God and the Devil as a single entity, each opposing half of which expresses an extreme of good or evil, conceivable in the human domain, then you will come close to sensing the way ancient Babylonians understood Ishtar. Her morning and evening celestial attitudes personified all the love and hate a human being could imagine rolled into one bright white light.

It is difficult to know when the Chaldeans took what appear to have been two distinct Venus goddesses and merged them into a single one with a dual personality. Maybe it happened when sky watchers recognized that the evening and morning star could be considered as different aspects of the same celestial figure. The visible descent and reascension so dominant in the Inanna folk tale evokes potent images. Evening Ishtar came to symbolize the goddess of love we traditionally associate with Venus. She was the power that attracts the sexes to each other—calm, peaceful, and placid—satisfying in her effect. But the morning after offers a different set of circumstances. Then a romantic affair can be perceived as merely lustful. The Ishtar of the dawn skies, who once smiled at the pleasures of love, now glared with abhorrence. Those who worshiped her in the morning would be disposed more to the love of war than flesh. So radically different was the meaning of morning and evening star light over ancient Chaldea.

Even when it came time to marry, Ishtar, alone among deities, was unsatisfied with a single spouse, the myth goes. She gave herself away on impulse to heroic men and gods alike. This behavior made her omens equally diverse; they portend war or famine on the one hand, fertility on the other.

> *Venus in the month* Tebet *rises* [?]. *Rain at sunset. The king establishes his hostile arms in the land. The troops* [march].
> *Venus in the month Sebat [makes] a rising* [nipkha] *in* sereti. *The crops of the land flourish.*[9]

Women under the influence of Ishtar were said to be unlike the Chaldean ideal, who lived in a harem and cared only to bear children. Rejecting the feminine role model, Ishtar pulled off her veil and descended into a life of debauchery. Herodotus tells of a custom in the Near East that existed almost up to the time he wrote his history (ca. fifth century B.C.): "Every woman born in the country must enter once during her lifetime the enclosure of the temple of Aphrodite, and must there sit down and unite herself to a stranger."[10]

Ishtar's Persian counterpart, Anahita, was the essence of fruitfulness. She purified the male seed, the female womb, and the milk in the breasts of all mothers. Strong, bright, and beautiful, she was adorned with gold, yet she was often portrayed as a seductive goddess, tightly girdled to emphasize her breasts, a symbol of the practice of temple prostitution (Figure 3-1c). In ancient Egypt, Venus became a two-headed falcon and carried the designation "The Star That Crosses," a possible reference to the side-to-side motion with respect to the sun that Venus makes on each appearance period.

If we backtrack far enough through Middle Eastern history, the early Venus does seem to become more androgynous. Under the Akkadian name of Dilbat, Venus was female at sunset but male at sunrise. The Moabite binary deity Ishtar-Kemosh was said to have once united the sexes, and Roman writers tell us that there was a statue of Venus on the island of Cyprus that had the body and dress of a woman but the beard and scepter of a man; there "they believe that she is both masculine and feminine."[11] And a reversal in the division of the sexes with male evening star (Tai-po) becoming female morning star (Nu Chien) existed in China,[12] except that Tai-po was supposed to be the husband of Nu Chien. Strange that the two are never seen together.

The Kédang of Indonesia also pair Venus as evening star (Uno) with Lia, the morning star. One of their Venus myths is called "The Morning Star Rises and Taps Palm Wine."[13] It tells of a villager who was angered at finding his palm wine containers empty when he retrieved them every morning from the top of his coconut tree. One night he waited in the treetop for the suspected culprit and caught Lia in the act of cutting off his supply source. He grabbed the sky god by his long beard but was dissuaded from pulling him down to earth by a promise that on the following night Lia would return with a surprise. Next night, thanks to the intervention of the brave tree climber, the mischievous god brought a golden pot of *uluq* (seed) to compensate for what he had taken away.

We do not know the details of the process, but Venus dualities from the Near East eventually got carried over the Aegean by Greek borrowers, who transformed them into terms that are more familiar to us. Phosphoros (or Eosphoros, after the Greek god of dawn, Eos), is the star of Aphrodite, the Greek equivalent of Ishtar, who sprang from the foam of the sea out of the castrated member of her father, Kronos— old father time. By contrast, the evening star image of Venus is associated with Hesperos, a male deity and the brother of Atlas, which becomes Vesper in Latin because this deputy of the moon prolonged daylight.

This dyadic myth, like so many of the Venus stories we have been talking about, really addresses the origin of the sexual principle. Where do genders come from? It seems to respond to the old chicken-egg problem by suggesting that just as Eve sprang from the rib of Adam, so procreativity was derived from a male property—what we might expect from these ancient, largely masculine accounts of the nature of creation. From this name we derive terms such as *phosphorescent* and *phosphoric* to connote a sudden shining as well as the name of the wax-colored element that ignites at low temperature. Lucifer (Luciferos), the later Latin equivalent, likewise means "bearer of light." (The ancestors of safety matches were once called Lucifers.)

Ideas, like trade goods and technologies, are passed along from culture to culture. The Athenians overcame the Persians; Roman imperialism later swallowed the Greeks—and the Middle East and north of Africa along with them—and when the Romans acquired the Acropolis, they also inherited a rich culture. No two cultures can come into contact without one influencing the beliefs of the other. The twofold aspect of Venus in the heavens was as resoundingly visible to the Romans as to the Greeks and Sumerians, although by the height of Imperial Rome the concrete imagery had been submerged in the myth, for the Romans, great warriors and administrators that they were, did not watch the sky very closely. The old sky stories became institutionalized, brought indoors out of contact with the world of natural experience where they once held more-immediate meaning. This is why Caesar never really perceived the concrete image of his patron deity.

How did the Roman name of Venus as morning star ultimately become linked to a term identified with Satan? The relationship between Venus and the sun highlights the opposition between the Prince of Darkness and the Christian God of Light and reminds us that Satan once was said to have fallen from God's kingdom.

> *If once he was as beautiful as he is ugly now,*
> *And lifted up his brows against his Maker,*
> *Well may all sorrow proceed from him.*[14]

To Dante the pilgrim, whom Virgil takes on a guided tour of the Inferno, the Satanic one is portrayed frozen upside down at the center of the earth, just where he landed when he fell from heaven and the grace of God.

Planetary dyads also existed in ancient China. Outer planets such as Mars and Jupiter were regarded as *yang* planets. They governed the centers of the Chinese city, while the inner planets—Mercury and Venus—were *yin* planets that looked after the peripheral domain—a rather odd dyadic switch to our way of thought. When *yin* and *yang* mingled, the planetary power was not dissolved or neutralized but rather combined to create a holistic and complementary kind of force. According to several omens passed down in the literature, the general doctrine called for a uniting of the military powers of the entire city—both ruling elite and surrounding subjects—against an archrival.

When Venus and Mars arrived together in the Pleiades, the Chinese, ever military and bureaucratic, foresaw unity in military combat. In Greek myth, the Pleiades represent Hephaistos's net, which captured Aphrodite (Venus) and Ares (Mars) in all the guilty pleasures of a carnal affair. Thus, the combination of these two actors in the same place on the celestial stage served as a metaphor for both love and war.

Alliances and heavenly power shifts appear to have been quite common among archaic deities and their manifest planets. Evidently there was as much political intrigue in the sky as here on earth. But some associations were far more complex than the simple pairs I have just described. Sometimes kings and cities sought their identities in a multitude of nature's gods. The Egyptian pantheon was rife with them. There were divine father, mother, and son, or a king with two wives; or, in the event that a goddess reigned over a home, she might have two or more husbands. For example, Sokhit, feudal goddess of Letopolis, and Bastit (from the city of Birbastes) became the wives of Ptah of Memphis in a triadic alliance that relied upon the proximity and economic connection among these cities and the interrelationships among their gods. Celestial appeal was an obvious way to cement ties and quell hostility among rivals.

Given the similarities we know about the ways they move, we can understand why Jupiter, Saturn, and Mars together were regarded by the Egyptians as the three Horuses, each corresponding to an aspect of their falcon-headed sky god. In an animate universe one understands nature by imparting to it all that happens in human life. Take the problem of the transfer of royal power. Just as one pharaoh succeeded another on earth, so too in heaven did Horus succeed his father, Osiris, who like all real kings ultimately disappeared from the real world to rule over the dead and eternity. Horus came in many tangible forms. He rose just as the moon and Nile waters rise. His falcon eyes were the sun and moon, and the wind became his breath. In planetary guise Jupiter was the Horus who illuminates the two lands (Upper and Lower Egypt), Saturn was Horus the Bull, and Mars the Red Horus or Horus of the Horizon.

Venus often figured prominently in the name sets of some of these Old World triads, usually together with the sun and moon. It is easy to see why if we recall some of the real harmonic beats in her celestial periodicities with those of the two more luminous deities. A trio from ancient Chaldea is composed of Sin (moon), Shamash (sun), and Ishtar (Venus) (Figure 3–1a). According to one myth, Ea, the earth god, charged his three children with rulership of the army of heaven. Among them he apportioned the day, designating that Venus should become the evening star who precedes the appearance of the moon and the morning star who announces the appearance of the sun.

A myth told by Shipibo men from the Amazon basin also links Venus, this time as the mediator, in a sexual triad with the sun and moon.[15] The sun is masculine. Women plead to him for his fertilizing power so that they can bear children in order to repopulate the world after the great flood. Shipibo say the moon—also a male—is responsible for menstruation. Every month he opens and shuts his mouth to imitate the opening and closing of the female orifice during the bleeding cycle. Sometimes the Shipibo moon changes sex and Venus acts as a substitute, whereas at other times Venus is described as the daughter of the sun. Once again the phenomenal basis of these Venusian transformations is found in the obvious close associations between that planet and the sun combined with its lunarlike mutability, its coming and going at the horizon just like the crescent moon.

In the pre-Columbian Yucatán peninsula, a visible heavenly triad in the form of a great planetary conjunction was once used to celebrate the foundation event in the story of the creation of the Mayan world: the birth of the gods. The mid–eighth century was a time of great tumult in West Yucatán, with the rival cities of the rain forest Toniná and Yaxchilán vying with Palenque for supremacy. Shield Pacal, the old king of Palenque, had succumbed after six decades on the throne. His was a reign of triumph and success in battles with these and other neighboring city-states. But now he was gone, and the people, like the Egyptians, needed a "Horus principle"—some assurance that his son and successor, Chan Bahlum, then in his late forties, would be a god incarnate like his father. The history of the gods at Palenque lay written alongside family genealogy, all freshly carved in stucco on a tablet about to be installed in Chan Bahlum's first temple. The text told of the descent of the ruling dynasty from three gods born at Palenque nearly four thousand years before. Each deity had been born of a father and mother who came into being several hundred years earlier, shortly before the most recent cycle of creation, just after the world had been destroyed by floods.

Although each of the gods in the threesome appears in many different monumental Mayan inscriptions, all seem to be linked in one form or another with celestial deities relatable to Palenque's contemporary rulers through "sky events." These rites of dynastic legitimation appear as statements implying that the rulers celebrated the occasions when their planetary ancestors actually convened in the sky. In some instances, the Mayan kings even depicted themselves in stone statuary as one or another god of the triad.

Who were these celestial deities? We can be sure one of them, the second born, was the sun because he is directly named *Kin Ahau* (Sun Lord) on the tablet. He is represented either as a young personage with the *kin* hieroglyphic sign on his forehead or as an older man with *kin* on his arm or thigh. Firstborn, the namesake of his father, has features in common with the sun deity—the same facial shape and squarish eyes. His name phrase has been deciphered as *Hunahpu*, which is the same as the name of Venus the morning star in the *Popol Vuh* creation myth (it means "1 Ahau," the name of the day assigned to the celebration of the appearance of morning star–Venus by the ancient Maya). Recall that in the Quiché version of creation, which comes down to us from pre-Conquest times, the second born becomes the rising sun, who is preceded in the sky by his firstborn brother, morning-star Venus. This makes sense because Venus announces the coming of the dawn.

The identity of the missing member of the Palenque triad is more difficult to pin down. Recognized most commonly as the god of the manikin-scepter, this bizarre-looking male deity has an upturned snout, one of his feet takes the form of a sky serpent, and smoke issues from his forehead, perhaps from a cigar implanted there in an obsidian mirror or crystal. Who can this be?

Having identified solar elements in one deity of the triad and characteristics of Venus in another, we might be tempted to try to pair the third with the moon or perhaps another planet, say Mars or Jupiter. But as anthropologist Dennis Tedlock, a translator of the *Popol Vuh*, has advised, it may be more realistic to think of certain actions of deities corresponding with depictions of kinds of movements or changes in celestial lights rather than with individual lights themselves. From my earlier discussion of celestial repose followed by the resumption of movement, recall that the myth stresses not so much the *object* that participates as the *process*, or what sort of change is going on.

The stories seem contradictory, even though the prevailing version says that Venus and the sun are brothers, Venus being firstborn (and carrying the name of his sire) and the sun, second born. Could the Maya have considered the sun and moon as one entity with dual characteristics? In one manifestation, one of the twins is called the god of the night sun, out of view most of the time but showing himself once each month in the form of a full moon. In this scenario, his brother is the daytime sun. When they interact on earth in the *Popol Vuh* creation story, the older deity takes the lead, whereas in the underworld the interaction is reversed. So the identification of the last born of the triad, based on the glyphic and linguistic evidence, remains a mystery, and a strong argument for a celestial attribution has yet to be put forward.

Yet we do have a ringingly clear statement that a great celestial reunion of the three creators was celebrated at Palenque in historical times. It is carved in stucco and celebrated in architectural form in Palenque's Group of the Cross complex (Figure 3-2). Each member of the great second-generation Palenque Triad of gods from whom the rulers of Palenque claim to have descended is featured in one of the three buildings that constitute the assemblage, one of the most stately in all the ruins of Mayaland. The south-facing Temple of the Cross houses the date of the birth of Venus. It is so named because the sculpted tablet in its interior contains the image of a cross-shaped tree of life that early explorers once misunderstood as a precontact Christian symbol. The tablet also contains statements about how the rulers are linked to Venus. The unknown member of the triad has his undeciphered name enshrined in the Temple of the Foliated Cross, one hundred yards to the south, which faces west; the Temple of the Sun, which is aimed toward the rising sun at the winter solstice, incorporates the name of the sun god.

The visible celestial encounter must have been one of the most important events of the young king's life, and there are some rather specific clues in the inscriptions on the three temples that tell how he and his people celebrated the start of his rule on the anniversary of the birth of the gods. At least four different monumental texts in the Cross Group tell us that the date was the equivalent of July 20, 690, in our calendar and that the celebration lasted for three days. Three bright celestial lights actually converged in the sky over Palenque on that night—but not the ones we might have expected. As dusk fell, there in the royal planetary lineup lay Saturn to the east, Jupiter in the middle, and Mars to the west. All appeared within a span of sky a few moon disks wide, high in the constellation of Scorpio—a rare conjunction. As

FIGURE 3-2. THE BIRTH OF THE GODS AT PALENQUE
In the plaza between the buildings that make up the Group of the Cross at Palenque, Mayan citizens assembled to celebrate the great conjunction of planetary deities on July 20, 690. The planets acted out the last creation of the world, believed to have taken place in the fourth millennium B.C. (photo by author).

night progressed, a gibbous moon lit up the horizon and joined them. Later all four coursed together over the high ridge that fronts the three temples on the south. In the early-morning light, we can imagine a crowd assembled in the open plaza facing King Chan Bahlum's temples. There they watched the reunited divine trio who had given birth to their ancient ancestors set over the Temple of the Inscriptions, the tomb where Chan Bahlum's father recently had been laid to rest—a symbolic affirmation that the celestial power once endowed the father would pass on to his son. Perhaps the king was inspired by these events to adopt Jupiter as his patron planet and future guide star, for it was brightest of the three.

The triad statement written in hieroglyphs in the Group of the Cross means "the crossing triplet gods" or "the three gods joined who cross the sky," argues epigrapher Barbara MacLeod. Was this a Mayan attempt to single out the fact that these very three (Saturn, Jupiter, and Mars), like the three Horuses of Egypt, are the only ones who can pass all the way around the sky? "The play's the thing," said Shakespeare. And so we wonder: How can the sun and Venus in the creation myth suddenly be transformed into Mars and Saturn in the enacted celestial version? And what does Jupiter have in common with the serpent-footed god of the smoking mirror? Are these really sensible questions to ask?

More likely the message given to the people on the night of July 20, 690, was simply that a representative triplet of gods came together to validate the continuation of the old king's rulership in the person of his son. The last few lines of the *Popol Vuh* read: "And then the two boys ascended this way, here into the middle of the light; and they ascended straight on into the sky, and the sun belongs to one and the moon to the other."[16] In other words, the hero twins, the sun and Venus in the underworld, become the sun and the moon in the sky. Like the Babylonian celestial gods, the Mayan sky deities often radically changed attributes. In our rational attempts to pair planets with gods one to one, we may run the risk of missing the main message of celestial name changing.

RESURRECTION, APICULTURE, HUMAN CONCEPTION, AND OTHER UNLIKELY LIKENESSES

The Venus of highland Mexico was a male. Quetzalcoatl may have been a real person turned legendary hero, high priest of the Toltec city of Tula, forerunner of the Aztec capital of Tenochtitlán; at least the people were taught to believe they descended from those who had followed him into exile. When he died, Quetzalcoatl was metamorphosed into the radiant Venus, who was prophesied to return one day to resume his rule.

Like Ishtar's descent and return, the death and resurrection of Quetzalcoatl are visibly manifest every time Venus disappears in the western sky in the evening and reappears in the eastern sky in the morning. The whole legend, replete with celestial dramatization, was cleverly played up by the conquering Spaniards to transform the pagan hero into a messianic figure like the Roman Catholic Christ. It is well known that Cortés landed on the coast of Veracruz at a time when the return of the god was anticipated by the cyclic calendar. An old Spanish chronicle reads:

At the time when the planet [as the evening star] was visible in the sky Quetzalcoatl died. And when Quetzalcoatl was dead he was not seen for four days; they said that then he dwelt in the underworld and for four more days he was bone [that is, he was emaciated, he was weak]; not until eight days had passed did the great star appear, that is, as the morning star. They said that then Quetzalcoatl ascended the throne as god.[17]

There are parallels between the Christian theme of the Resurrection and the apotheosis of Quetzalcoatl that have nothing to do with the diffusion of Western religion. The movement of Venus, especially around the time of its disappearance and reappearance, is most evocative—an ideal celestial metaphor for resurrection viewed in any sky. Bernardino de Sahagún, Spanish chronicler of the Aztecs, tells how the people sacrificed to Venus once he came forth:

Of the morning star, the great star, it is said that when first it emerged and came forth, four times it vanished and disappeared quickly. And afterwards it burst forth completely, took its place in full light, became brilliant, and shown white. Like the moon's rays, so did it shine . . . [captives] were slain when it emerged, [that] it might be nourished. They sprinkled blood toward it. The blood of captives they spattered toward it, flipping the middle finger from the thumb; they cast [the blood] as an offering; they raised it in dedication.[18]

It is difficult to sort out the truth when it is told by a foreigner writing amid the infidel. Sahagún was a Roman Catholic priest, and most of his native informants were military men far from nature's fertile fields. Rarely were they experienced in natural history. He asked them questions such as What do you call this object or phenomenon (as he pointed to a figure on a piece of paper)? What is its nature? And he received answers such as: eclipses of the moon (our pregnant women are frightened of them) or eclipses of the sun (monsters descend). Unfortunately, such queries and their recorded responses tell us precious little about the empirical quality of planet watching and its relation to Aztec philosophy and religion. This European holy man, like so many others, seemed to dismiss Aztec astronomy, what little he perceived of it, along with all natural philosophy, as lowly and rather minimal.[19]

However, we know that the great Aztec imperial state ruled over most of what we now call Mexico for two centuries up to the eve of the Conquest. Its island capital of Tenochtitlán, today Mexico City, boasted over one hundred temples and a population close to a quarter million people. The Templo Mayor (Great Temple) was lined up to point to the sun when it rose on the equinox. At the top of the temple, priests sacrificed thousands of captives to keep the world running. Twin gods, one of rain and fertility, the other of sun and war, had their separate sanctuaries positioned there. According to Sahagún, blood sacrifice was related to the resurrection of the Venus deity, but in which temple these rites were performed we cannot say. Today little of the ancient capital has survived, and most of it lies submerged beneath the sewers and subway tunnels of modern Mexico City.

The Codex Chimalpopoca, another book written just after the Conquest, tells us that the priests knew precisely when Quetzalcoatl-Venus would appear. Then he would cast down his rays upon all to show his displeasure. But it would be decided whom he would spear or cast omens upon with his dazzling rays (just as in the Fertile Crescent) by the day he first appeared. If the day was, for instance, 1 Cipactli,* then old people were affected, but if it was 1 Ollin (Motion), it would be the young. If it was 1 Ocelotl (Jaguar), 1 Mazatl (Deer), or 1 Xochitl (Flower), children were speared; if it was 1 Acatl (Reed) or Miquiztli (Death), the lords; if on 1 Quiauitl (Rain), the rain (in other words, it would not rain). If 1 Atl (Water), then everything would dry up.[20]

We find this same selective equation between appearance and omen in pictorial manuscripts all over Mesoamerica long before Cortés and his men came to American shores. Sacred books show victims being speared by Quetzalcoatl-Kukulcan in various guises; for example, in the Dresden Venus Table as well as in the Cospi Codex from Central Mexico (Figure 3-3), we find five different countenances and costumes represented by the so-called Great Star. Each Venus deity hurls his omen-bearing darts of affliction at humankind, one for each of the five courses Venus takes across the sky. Take a closer look, for example, at the Dresden Table (Figure 4-1), so named after the city in whose library the document turned up over three centuries after it left Mexico. This interesting table will occupy considerable attention in the next chapter, when I explore some of the precise, mathematically expressible aspects of early planet watching.

On page 47 of the Dresden Codex (second frame of the sequence), Venus is Lahun Chan, the god of the tenth heaven (there were thirteen in all), and he wears the Venus symbol in his headdress. He may have been a form of the Venus god of the Maya before Quetzalcoatl was imported from Central Mexico to take his place.[21] He then possessed the head of a jaguar and the body of a dog and was said to have been envious, dishonest—a sower of discord—a god without virtue; but his jaguar teeth and claws were powerful and frightening.[22] This regent guards the west, where all things go to die, which is why his head is

---

* The 1 refers to one of a set of numbers 1–13 and Cipactli is the name of one of the 20 days of the Aztec week. Like the division of the zodiac, most days are named after plants and animals, Cipactli being Alligator. Just as 30 numbers and 7 days of the week may be paired successively in our calendar, so the two Aztec time wheels of 13 and 20 roll round and round together, forming a ritual cycle of 260 days.

fleshless and his skeletal ribs protrude. He represents Venus in his western or evening star aspect.

By contrast, the Venus who does the spearing in the fifth frame of the Dresden Codex has been related to the Central Mexican god Tlahuizcalpantecuhtli, who in certain guises can be the god of frost as well as the Lord of the Dawn. They say he once shot an arrow at the sun to get him moving, but he missed and the sun shot back, hitting Tlahuizcalpantecuhtli and throwing him face down into the underworld. Thus did he become the god of cold, says the Codex Chimalpopoca.[23] Here is an obvious reference to morning star Venus during the cold periods when dawn and frost happen together, when the planet has just reemerged from the underworld, where it spends all its time while absent from the sky.

While Greek and Roman art pays attention to how you look, Mesoamerican imagery seems to focus more on what you wear. These New World mythic figures seem more variegated than their Greek or Roman counterparts, indeed as different as each costume that adorns the Venus god in the Dresden codex. They masquerade and impersonate. Individually they seem less rigidly assignable to particular qualities of nature, but if we look carefully we can discover likenesses their human creators perceived hidden away in the texts they have left behind for us to decipher.

Take the table used to predict lunar and solar eclipses immediately adjacent to the five Venus gods in the Dresden Codex. On the last page is, of all images, a hieroglyph representing the planet Venus (see Figure 3-4e), so identified because of its repeated appearance on the pages of Venus calculations that precede it. The glyph serves as the head of a "diving god," a male figure shown plummeting from a sky band—the segmented body of a great celestial serpent. But what is a Venus god doing in an eclipse table, his inverted feet pressed against a pair of hieroglyphs that represent eclipses—half-light half-dark fields on which the signs for day and night are superimposed? Stranger still, of all ten pictures in this section of the manuscript that represent eclipses, only the one with the Venus-headed god is shown upside down in the act of descending.

Notice his beelike abdomen and a sharp protruding stinger. This diving god in beelike form is well known all over the east coast of the Yucatán peninsula. Half a millennium before Columbus, the Maya sculpted it over the doorways of ceremonial buildings. The ancient Maya revered bees, admiring the community life of the hive as a social role model. The gods who upheld the world at each of its four corners were patrons of the bee community. Like all gods, they were sustained by *balche*, the intoxicating brew made from bees' honey. Today anyone who visits the small villages surrounding the tourist zones of the ruins of Cobá and Tulum near the island of Cancún will find that beekeeping

FIGURE 3-3. THE FIVE COSTUMES OF THE CENTRAL MEXICAN MALE VENUS, ONE FOR EACH UNIQUE COURSE OF THE PLANET IN THE SKY.

Below each deity is his impaled victim, each a metaphor for an omen. There are also five Venus deities in Mayaland, and they perform the same function (compare Figure 4-1). Sources: (Graz: Akad. Druck–Ü Verlag)

3-4a.

3-4b.

3-4c.

FIGURE 3-4. WRITTEN FORMS OF THE MAYAN NAME OF VENUS.

a. Venus resting, a possible Mayan reference to the time when the planet lies at its greatest height above the horizon.

b. The standard form of the Venus glyph in Mayan codices is a quincunx made up of four dots surrounding a point at the center.

c. Sometimes the image is halved as painted on the wall of the Palace Tower at Palenque.

3-4d.

d. Often it appears as part of a spoken phrase, such as "star-over-earth" (Drawings by Michael Closs, photo by author)

3-4e. The Venus-eclipse-bee diving god in the Mayan Dresden Codex. Likenesses our eyes never could have perceived? (Source: Villacorta and Villacorta)

3-4f. In highland Mexico one form of the pictured name of Venus may have looked like a crescent moon, perhaps seen here rising in the doorway of a temple. Was the crescent witnessed before the advent of the telescope?
(Source: Graz: Akad. Druck–Ü Verlag)

3-4g. From an ancient Mesopotamian tablet (No. 65) from the City of Uruk, Venus is depicted as a star. The inscription suggests that Inanna follows the setting sun into the underworld (Source: A. Falkenstein)

and the production of honey still flourish. In one Mayan dialect, Venus the morning star is called Xux Ek or wasp star (although it is not clear how wasps and bees were related in the Mayan taxonomy).

What do bees and Venus have in common, and what do they have to do with eclipse predictions and the deity in the descending posture? In the early sixties, a pair of Australian zoologists performed an interesting experiment with bees that may provide some answers. Their studies dealt specifically with the habit of departure from and reentry into the hive of certain species of stingless bees, near relatives of the Yucatec variety. Knowing that bees navigate according to the position of the sun, the zoologists wondered how their movement might be dependent on the time of day and the season of the year in the tropics, where the noontime position of the sun can migrate to the north as well as the south of the zenith or overhead position. They conducted their experiments at three tropical latitudes.

To their surprise, the researchers found that during that part of the year when the noonday sun lay south of the zenith, the bees entered and left home base in a clockwise direction. But when the sun stood to the north, they suddenly reversed course and moved in a counterclockwise fashion. Now, in tropical latitudes there are only two days of the year when the sun stands directly overhead at noon, after which it begins to descend either to the north or to the south. The dates of solar zenith passage represent a calendrical line of demarcation between the seasons of the northern and southern solar course. Precisely when this happens depends on the latitude of the site. In all three instances in the bee experiment, the reversal of the bees' course happened on or about the zenith passage day at each location. Given that the bees take their navigational cues from the angle of the sun, this result ought not be surprising.

Is it possible that Mayan beekeepers connected the direction of circulation of bees with the place of the sun in the sky a thousand years before modern biologists? If so, the icon of the descending bee god may have been one way the Maya chose to express their knowledge and appreciation of the perceived behavior of one of their precious living resources, upon whose nourishing royal nectar they depended for their survival.

This hidden likeness between bees and the sun might be evident to anyone who spends a good deal of time observing bees. But where does the Venus symbolism fit in? Imagine a throng of people assembled in the great open plazas of ancient Tulum by the sea. Obscure during the Classical Mayan period, by the ninth century the city had grown into a major port, one of the centers of the salt trade that burgeoned along Yucatán's east coast. Less interested in the elaborate hieroglyphs, calendar keeping, and monumental architecture of their predecessors, the East Coast Yucatec Maya of the Late Postclassic period seem rather more commerce oriented. Nevertheless they continued to worship sky deities, particularly the diving god—and the sculpture adorning the cornices of many of Tulum's buildings shows it.

The date is June 26, 884, shortly before noon, and the dark new moon has just bitten into the disk of the sun—the start of a total eclipse in Yucatán. Spectators watch night slowly descend upon them. With the intervening darkness, the humid air is suddenly cooled by a breeze off the Caribbean. Gradually the sea alters in color, light green to azure, a white foam appearing beyond the surf. Animals scurry confusedly about the landscape, and birds scatter in flocks overhead. As the thin, bright solar crescent gives way to the diamond-ring flash of the last gleam of sunlight—there, appearing just off the western edge of the sun, is bright Venus, its rays plunging earthward along with the darkness. The planet actually had vanished from the nighttime sky only a few days before, and it was not scheduled to reappear for at least another month. The astronomers had predicted it all with their codex. But today the people were offered a rare glimpse of their descending deity while he lay in hiding.

Why the Venus-glyph-headed god was placed at this particular juncture of the Dresden eclipse prediction table, we cannot really say. No Mayan scholar has been able to match up the pictures in the table with a chain of actual eclipses in a convincing way; nonetheless, we can be certain the document was used as a warning device to anticipate these calamitous omen-bearing events: "Woe to the corn crop, woe to the pregnant female, the pestilence brought down by the sky gods is upon us," reads the codex. A few eclipses were visible in Yucatán during the ninth and tenth centuries. I chose a fairly spectacular one that happened at a pivotal time in the heyday of Tulum—at midday close to the June solstice. Recall that this is the time of year when the sun reaches its northernmost excursion point, turns around, and begins its trek back toward the south. Also it was a time when Venus was in one of the brief disappearance portions of its four-phase cycle—it happened to lie close to the disk of the sun. Mayan hieroglyphic books are about more than predicting eclipses or planetary apparitions. They tell us that the Maya had looked with great diligence at the subtle interplay among seemingly diverse elements of nature and expressed their discoveries in ways that seem entirely alien to both our eyes and our minds.

The sun at its zenith, daggers of Venusian light, and the darkness of the eclipse all share a single quality: They descend to earth. The descent theme may be repeated in a form of the Venus hieroglyph (Figure 3-4a) that shows the figure of a man seated with hands clasped over his drawn-up knees, his head resting upon them. The date that accompanies the glyph corresponds to the time when Venus lies at its greatest distance from the sun. Recall from Chapter 2's naked-eye astronomy lesson that this is one of those times in the Venus cycle when any movement of the planet is particularly hard to detect. It hangs motionless in the evening sky from night to night, relative to sun and earth. Like a celestial roller coaster, Venus's downward slide, at first barely perceptible, gradually develops into a rapid plunge. In Figure 3-4a, argues epigrapher Michael Closs, the scribe has portrayed Venus at rest, before it resumes its inevitable descent, imaginatively and thoughtfully capturing his essence in the curious character with the bowed head.

Mayan astronomers seemed quite interested in sky representations of this idea of halting before taking precipitous action, like the calm contemplation before waging a battle or entering into an alliance. Another like-in-kind observation marks the time when Jupiter halts and then resumes its normal west-to-east movement among the stars—the end of its retrograde motion.

I mentioned earlier that Chan Bahlum (the name means "serpent-jaguar"), early eighth-century ruler of Palenque in Mexico, adopted Jupiter as his patron planet. On several monuments dedicated to him, the king is shown dancing with underworld figures as he undergoes his apotheosis. Dates accompanying the sculpted panels in his temple refer to key events in his life; for example, the date he was officially designated as heir to his father's throne—he was only six at the time—the date he acceded to rule, the date he took the throne at the age of forty-nine, the eighth anniversary of his accession, and the commemoration of his death five months after he died at the age of sixty-seven. All of these dates fall close to the time Jupiter makes its first perceptible departure after retrograde, as he embarks on his west-east celestial journey. But although an accession date and even the celebration of a ruler's entry into the underworld can be scheduled to fall on Jovian departure dates, how could such celestial timing be achieved with an individual's birth? Clearly in some instances the Maya must have altered history's dates after the fact.

If Venus and bees seem disjointed from each other in our world, consider Venus and human conception, in which the Maya also found unity. They did so by discovering an unusual coincidence. It is well known in our culture that the birth cycle can be related to the phases of the moon. What pregnant woman does not count her term by simply flipping through nine pages of the calendar? But Venus?

One account from a Mexican chronicle reads:

> In this land the star [Venus] lingers and rises in the east as many days as in the west—that is to say, for another period of 260 days. . . . They also kept account, like good astrologers, of all the days when the star was visible. The reason why this star was held in such esteem by the lords and people, and the reason why they counted the days by this star and yielded reverence and offered sacrifices to it, was because these deluded natives thought or

believed that when one of their principal gods, named Topiltzin or Quetzal-coatl, died and left this world, he transformed himself into that resplendent star.[24]

We know that this 260-day cycle or count of the days, called the *tzolkin* by the Maya and *tonalpohualli* by the Aztecs, lay at the core of all Mesoamerican calendars, at least since the sixth century B.C. This cycle rose to prominence, I think, because it approximated the length of several fundamental life-sustaining periods: It was a measure of the duration of the agricultural season, nine lunar months (266 days), as well as the Venus appearance interval mentioned in the preceding quotations, and it was equal to 13×20, both sacred numbers.* In the remote Mayan highland region, where the modern calendar has not completely overtaken the pre-Conquest one, anthropologists have recently discovered that women still associate the *tzolkin* with the human gestation period, which modern biologists estimate at between 255 and 266 days. Biorhythmic cycles and their connection with celestial periodicities need to be examined more closely by people who wish to study native calendars and timekeeping. These simple practical intervals serve as the ultimate model out of which more complex, artificial time intervals are contrived by the rational mind.

Unfortunately we know far less about planetary symbols embedded in the myths of cultures that thrived in other parts of the world before the sudden European intrusion in the age of discovery and exploration. But what little we can learn from early contact stories is enough to demonstrate that people used the sky as a medium to direct everyday occurrences that especially concerned them.

For the ocean-bound people of ancient Hawaii, the planets were sustainers, supporters or pillars of a giant celestial round house built on the model of the houses in which they once lived. They were placed in the sky to help people—to warn them of coming events. To learn the warning system, one only needed to pay close attention and follow the planets' movement among the stars from year to year. Some planetary deities paid special attention to the fishermen, others to the tattoo artists—we cannot explain why in every case. Like the stars, the planets could intermarry and breed children. Venus, as we might expect, was by far the most prominent. It was variously known as Dog of the Morning, Star of Day, and Forerunner of the Morning, a status similar to that of the Greco-Roman Phosphoros-Lucifer.

When Hawaiian Venus was the Great Star, its name was Hoku-loa, which placed it in the same class as the sun and moon, but when it

---

* There were thirteen levels in the Mayan heaven, and 20 as the base of the Mayan mathematical system no doubt originated from the number of fingers and toes, with which these people tallied before the advent of writing.

dodged unsteadily from side to side, like Mercury, it was dubbed the Royal Inebriate. For seafarers like the Hawaiians, weather prediction ranked high in importance alongside astrological affairs of war and state. Naholoholo, literally "swift-moving," like the storms in these parts, is one old Hawaiian name given to Venus. In one tale, the Great Guide Star of the Evening deviates from his course to suppress the fury of a hurricane and, as a result, loses his balance and falls out of his canoe.[25]

Venus and the wind live up to this mythic attachment the world over, for if watched carefully Venus can indeed be a predictor of wind as well as rain. For the ancient Mexicans, Ehécatl, god of the wind, was one form of Quetzalcoatl, the Venus god. Round temples, which imitate the shape of the dust tornado, were dedicated to his worship. In Greek mythology Dawn (Eos) and Morning Star–Venus (Eosphoros), to whom she gave birth, are related to the winds. Is this just because the wind kicks up at dawn? In the *Theogony*, a Greek creation myth, eighth-century B.C. poet Hesiod tells us that the three winds are also associated with seasons, the North Wind (Boreas) with winter, the West Wind (Zephyros) with summer, and the South Wind (Notos) with autumn. (The Greeks recognized only three seasons.) In his poem *Works and Days*, Hesiod ties the arrival of these winds to specific astral references; each has a time and place in the calendar determined by the wheeling motion of the celestial sphere. For example, when the bright star Sirius perches over our heads, we turn our faces west to enjoy the cool breezes of the Zephyr. This poetic language was part of the dialect for expressing the way the Greeks understood nature.

Like the Maya, the Maori of New Zealand used conjunctions of planets as an effective means of casting war omens. A conjunction of Venus and the moon determined whether a siege would be successful, and where Venus lay relative to Mercury, to the right or left, connoted success or failure in battle. In Tahiti, when Venus and Jupiter, the two brightest planets, set together near the horizon, it meant that two chiefs were conspiring against each other.

The Skidi Pawnee of Kansas and Oklahoma were among the most avid pre-Columbian planet watchers. They followed the continuing drama being acted out by their celestial deities, the sun and moon, but were especially attentive to the planets. Planetary movements mimicked their kinship relations. When in conjunction, they were brothers who visited one another. They also undertook general celestial duties, such as housing the souls of those who died from disease. In return for their services, villagers offered them smoke. As recently as 1838, they were said to have offered a human sacrifice to Venus,[26] a payment they believed they owed the Great Star because of his primary role in creating humans. Pawnee myth says that we are all born out of a battle of the sexes, a conflict involving Morning Star (male) of the east, who presided over the celestial council and favored the creation, and Evening Star (female) of the west, who opposed it. All the male stars of the east went forth to court her. In fact, we still see evidence of the courtship in the east-west motion of all objects in the sky. Morning Star prevailed and persuaded Evening Star into sexual union. Thus came forth their first child, who was a woman.

It is interesting that in many cultures the further back we trace creation myths, the more hazy they become about whether man or woman comes first in the genealogy. Perhaps this only reflects the magnitude of the problem about where the principle of sexual union originates. For example, in the Babylonian *Enuma Elish*, the blending of the male and female principle is manifested in the commingling of fresh and salt waters where the Tigris and Euphrates meet the Persian Gulf. Before this time the universe was a watery chaos not unlike the chaos in biblical Genesis. Then all things, earth and sky, land and sea, light and dark were divided by the hand of God. Ometecuhtli, the creation deity of the Aztecs, also is bisexual, man and woman at the same time, and he/she resides in the uppermost layer of heaven. The celestial dualisms I have been describing—sun and moon, evening and morning star Venus—are visible reminders of ideas and principles conjured up in the human mind. They seem to speak to all of us about the order in nature—telling us that so much of the real world seems to be resolvable into dipolar yet complementary opposites.

SKY MIRRORS LIFE

The lesson of this chapter has been that the words that make up sky stories can be misleading as well as revealing. For us the word *myth* has come to mean a story dreamed up out of pure conjecture—having little basis in fact. Taken to the extreme it is a fabrication, totally unjustified when held against real-life experience, as some would say of the myth of the American Dream, the Loch Ness monster, Babe Ruth, or Atlantis. But the astral myths examined in this chapter, some in more detail than others, seem deeply ingrained with truth, even for those who do not tell them. The very names given Mars, Jupiter, and Venus by Sumerian, Mayan, and Hawaiian peoples describe the way these celestial deities actually behave as they move along their courses. These strange stories once evoked real images of the planets to which our modern culture has payed little attention—grouping Venus's disappearance in fives, tying its motion to the coming of rain and wind, to human gestation, to bees entering and leaving the hive. Myths *do* have a basis in fact, even though the facts are not categorized and grouped in a way familiar to us.

We all seek unity, but the way people create order in the world around them seems to depend very heavily on their social agenda—what they need to function in everyday life—for it is culture that gives nature its structure. The hidden links between the way the wind blows, the stars move, the dawn breaks, appear to differ the world over, and the process of bringing disparate parts of the lived environment together is not shared by all people—except at the most basic levels. What happens overhead is not some set of abstract cultural universal principles to be comprehended in the same way by all of the people all of the time. We are too diverse for that.

This chapter's exploration of names traced through sky stories has led to the revelation of different kinds of common sense. For example, our ingrained way of thinking teaches us to pigeonhole observations of nature in distinct disciplinary categories. A ring around the moon has nothing to do with the moon, we say. Rather it portends a permutation in the atmosphere: "Rain is coming soon." Changes in the air are not affected by the moon, we say. It is only the light being refracted by ice crystals that betrays the atmosphere's moisture-laden presence. Likewise the colored halo can have no influence on the moon—upon its mountains, craters, or any Apollo landing craft that still remain on its surface. Unlike many of our predecessors, we neither propose nor even contemplate an overarching theory to envelop meteorological and astronomical changes; therefore we do not look for such connections in the explanations offered by other cultures. But, as we shall see in Chapter 6, it took a long time to embed this compartmentalized view of nature into the grain of Western common sense. Indeed, the process was completed only comparatively recently. If we fail to take seriously the stories of Venus and bees, Jupiter and kings, that originate from other common senses, we run the risk of forcing our forebears into the mold of our own contemporary brand of common sense, and this deprives us not only of knowing them but also of fully appreciating how we have become who we are.

Star myths may strike us as amusing stories, but behind the planetary alliances we have examined lie real people asking the kinds of questions we no longer ask the sky: What is the origin of gender and sex? Where does fertility—or for that matter any power—come from? Where do we go when we die? How can we foretell the future? And, as I have shown, many of their inquiries can be framed in the dialogue of visible planetary characteristics and changes: descent and resurrection (particularly for Mercury and Venus), dyadic and triadic bonds. No wonder all these ideas are so prominent in the early sky mythologies that grew up in both the Old World and the New.

Which came first, the myth or the sky observation? The scientist or the cosmos? No one can really say, but I think most would agree that the motion of things in the sky surely must have served as a very early practical timekeeping device, at least for those cultures that decided to look upward. Naming the phases of the moon and associating the course of the sun across the zodiac with seasonal activities date back into history as far as any document can reach. It would have been logical to marry the act of storytelling about everyday affairs to acts of nature simply as a way to embellish and lend structure to time—to remember how to mark its repeatable cycles. And with the process of storytelling came the logical expansion into more fundamental and speculative questions: Where did we come from and what will happen to us in the future? In some instances the forces of speculation, together with the need, especially in highly structured societies, to formalize in detail the relationship between people and the sky, led to some rather extraordinary myths.

In the next chapter I move, just as some of our ancestors did, from qualitatively describing, naming, and invoking the moving image in myth to exploring the ultimate rigor that often underlies the act of following it—the art of planetary prediction. This will lead to an exploration of some of the reasons behind the curious obsession with precision and mathematical exactitude that, in one instance we know quite well, led to the foundation of modern scientific astronomy.

CHAPTER 4

ASTRONOMY:
FOLLOWING THE IMAGES

I am convinced that real progress in the study of the history of science requires the highest specialization. In contrast to the usual lamentation, I believe that only the most intimate knowledge of details reveals some traces of the overwhelming richness of the processes of intellectual life.

—OTTO NEUGEBAUER, 1941, P. 13

PROCEED WITH CAUTION

Scientists delight in discussing the histories of their disciplines. Biologists admire Aristotle's careful description of the life cycle of a chick embryo, chemists the elegant construction of the periodic table of the elements, and astronomers the descriptions by their Greek counterparts of how the sun, moon, and stars turn about the sky. In these simple classification schemes, experiments, and rudimentary applications of mathematics and rational logic, we discover the roots of our modern way of knowing nature. But mention alchemy to a chemist or metallurgist, divination by hepatoscopy* to the medical researcher, augury to the ornithologist, or astrology to the astronomer and you will

---

* Hepatoscopy, or examining parts of the liver to tell the future, was so common a practice as to be licensed in ancient Greece. The largest organ in the body and the center of life was thought to reflect the state of the universe at the moment of sacrifice of an animal offered to the gods. But at a rational level an examination of the material processed by the livers of animals who grazed on a potential settlement site could tell a lot about the degree to which the health of an ambient community would profit from the local environmental conditions (for details, see J. Rykwert, *The Idea of a Town*, 41–54). Augury, or divining by observing the chatter or flight of birds, also was popular in ancient Greece. Today the word has survived and means simply "a foretelling."

likely be met with derision. Why? Because today we see these endeavors as wayward paths that detoured the human intellect on its evolutionary course of progress. Such occult practices are the rattling skeletons in the closet of modern science—aspects of our past worth shutting and bolting the door on because they do not belong on the upward trajectory of development we believe has led to our present understanding—the correct way to pattern the universe.

But this so-called nonscientific rubble we leave by the wayside can be revealing, for when we pick up the bits and pieces and study them closely, as historian of science Otto Neugebauer recommends in the epigraph I have chosen for this chapter, they offer up rich detail on alternative ways of thinking about the world.

In the previous chapter I showed that sky watching served everyday needs and concerns. Brought together under the umbrella of hope and desire, people wove celestial imagery with ideas culled from their imaginations to create fantastic tales called myths that were told around the campfire—stories concerned with how the human condition came about. In the present chapter I am going to demonstrate that in many cases myth passed well beyond basic function. Framed in an often elaborate mathematical context, its capacity to answer basic questions sometimes became more precise than it needed to be to keep society operating in an orderly manner. As I continue to explore the intersecting worlds of myth and science by examining ancient rudimentary forms of astronomy, it will become a bit easier to see these mythmakers as we see ourselves. The measure of mathematical precision I am going to unveil may surprise some readers, especially when they realize that, although we can label some of our forebears' considerations scientific, such efforts nonetheless served religious needs.

Let the reader beware. I am about to delve into some of the technical rigor behind the planetary imagery discussed earlier. Again the epigraph tells why. I shall explore some of the intricate celestial rhythms in an unfamiliar key, but such scrutiny offers rewards well worth attaining.

Take our modern calendar—a bundle of days stacked into weeks, all tucked neatly onto a twelve-page monthly musical score written in the key of "sun"—an annual packet of time. When we discard that list, around the time we make our New Year's resolutions, we set ourselves again upon the course of the sun, whose temporal signposts are marked by the cycle of the seasons as the earth makes its voyage around it. The sun has become the basic meter for charting both human and natural events. Now imagine instead the calendar of a fictitious culture that used the moon's cycle as a base. For them, planting and harvesting, hunting and gathering, even the course of the other lights that wander across the heavens, would march to a different beat. If such people composed their calendar, its musical score might sound very dissonant to us. But what would it look like? Could we even recognize it? We need not resort to science fiction to find out. In the Islamic world, the religious calendar is still basically lunar. In fact, our own academic time word *semester*, originally a period of six months, may be connected with the cycle of eclipses, and it has a decidedly lunar origin.

To make the point—that myth and astronomy harmonize in a strange way—I am going to analyze in depth two celestial scores, written texts from cultures on opposite sides of the ancient world. One is said to have contributed substantially to the foundations of our own form of modern science, the other grew in total isolation from the Western world.

In the bark-painted Dresden Codex, written in Yucatán three hundred years before the Spanish invasion, about A.D. 1200, Mayan astronomers manipulated the observed motion of Venus and canonized it to march to a lunar beat in exactly the way we have maneuvered the calendar year marked by the course of the sun to keep it in tune with the seasons. The Maya deliberately distorted the intervals of Venus's presence in the morning and evening skies discussed in Chapter 1 to fit segments and multiples of the cycle of moon phases. Yet they were able to keep track of precisely when Venus would appear or disappear.

The second example, from seventeenth-century B.C. Babylonia, is popularly known as the Venus Tablet of King Ammizaduga. Its contents, carved in clay, bear a remarkable likeness to the information in the Mayan document, and scholars are hard-pressed to explain how cultures so remote that they never could have been in contact could create such intricate, nearly identical messages. Except for the day, the shortest, most easily recognizable time frame based on sky watching is the month. As the all-sky drawing in Figure 2-2 indicates, we can follow the moon through its full cycle of phases by looking in the west after sunset or in the east before sunrise. We quarter our months into weeks because we can see the division by seven-day intervals, for example, between half and full moons. Preliterate people who interact in the open environment would find it easy to say, "We will meet again when the moon looks like this" (gesturing its shape), or "two moons from now" (indicating the count with the fingers), or "when it appears in the celestial house of [one of the constellations of the zodiac]." Such visible reckonings would be convenient for measuring out the cycle of raising and selling crops or the duration of hunting and fishing trips. They provide a ready celestial response to simple human needs. Little wonder then that most early calendars are constructed about a lunar baseline rather than the more lengthy solar one. Still, as anyone who tends a garden year-round knows, plants as well as animal life respond to the seasons. Therefore, as I delve into ancient calendrical documents, lunar cycles will be seen to mesh with the year. Many hunter-gatherer as well as agrarian cultures give specific "activity" names to the twelve or thirteen full moons that fit into the seasonal year (Harvest Moon and Hunter's Moon are two that survive in North America).

Ancient Middle Eastern planet watchers knew that if Venus was first visible in the morning sky after its eight-day disappearance on the day of a full moon, then the same celestial crossing would occur eight years later. And so, for eight of the twenty-one years of the reign of their king, Ammizaduga, court astronomers dutifully listed the appearances and disappearances of Venus—the ones that appear in his Venus Tablet—in a lunar-based notation. And like the Maya, who also breathed an air of astrology mixed with astronomy in their Dresden Codex, the Babylonians believed that each separate lunar month carried its own omen.

When we try to view a painting or listen to a concerto "objectively," we are automatically influenced by past impressions—what we think we know of the artist, when the composer lived, or what current taste tells us about what we ought to appreciate or disregard. And when we confront the unknown, we cannot fail to draw on what we already know, as Jacob Bronowski says. In Chapter 2, I tried to look beyond the way we have come to understand the planets as a class of objects that move about the sun in a vast space-bound universe. I constructed the perspective of an earth-centered viewer with limited technology. Now I am going to show how some of the perceptions described there and explored in sky myths in Chapter 3 were in fact well integrated into the more scientific astronomies of ancient cultures; for example, the disappearance intervals of Venus before first rise in the morning sky and its five-to-eight commensurability with the solar year were recorded in detail by Babylonians and Maya alike.

Because their messages were never intended for us, we need to approach decoding ancient astronomical texts the way we enter a city street from one of its busiest intersections. Cautiously, we pull up to the stop sign and look both ways—first to the material evidence, then to the appearance of the sky. We shift our eyes back and forth attentively, and with each glance we build up a clearer idea of when and how to proceed. To begin with, we need to appreciate in the most general way how these ancient calendars really worked. They were created to chart out patterns of regular, repeatable phenomena, such as Venus appearances and moon phases and solar positions on the horizon—not extraordinary or cataclysmic phenomena, such as supernova explosions or spectacular comets—the sky events that tend to captivate our modern interest. Knowledge of intricate patterns is important in societies whose existence depends upon the slight deviations from normal that can be anticipated in nature by those acquainted with its basic form—an extraordinary high tide, the ultrahigh cresting of the river, heavy rainfall or drought. What fascinates me about the Mayan codex and Babylonian tablet I am about to probe is the degree to which the calendar priests were absorbed with nature's detail.

87

KUKULCAN and IXCHEL

The Maya are recognized as one of the foremost civilizations of the New World. To judge by the remains, their origins were part Olmec (from the Gulf Coast south of Veracruz) and part Zapotec (from the southern highlands of Mexico). By the third century B.C. they had erected vast complexes such as El Mirador, deep in the rain forest of northern Guatemala, and Edzna, near the west coast of Yucatán. Here monumental pyramids, public buildings, reservoirs, and causeways demonstrate that the first true Mayan states consisted of ten thousand or more people residing in splendid cities, individual major centers around which smaller dependent villages grew. Archeologists have unearthed tools of both farmer and artisan. Elite burials and a handful of monumental inscriptions testify that America's first cities were highly centralized and specialized habitats. Separate complexes of palaces and temples in different parts of the urban environment seem to have been organized around their own central plazas connected by paved roadways. This suggests a division of power, likely based on kinship among the elite classes.

In addition to political and civic transformations, sweeping religious change also seems to have affected the Maya at about the beginning of our Christian era. It is manifested in messages about the gods and their relation to real people in works of art sculpted on cornices and over doorways of religious structures as well as in the layout and orientation of special buildings toward the cosmic directions. Later dynasties left behind lengthy inscriptions testifying to the cosmic power of Mayan leaders. Stormy Sky of the eighth cycle of Tikal (early second century A.D.), the eleventh ruler in his blood line, helped establish his city's economic primacy throughout the heart of the Yucatán peninsula. Smoke Jaguar, late in the ninth cycle of Copán (early sixth century A.D.), was a king even in the European sense of that word. He had a small empire consisting of a handful of surrounding sites and was remembered in Copán inscriptions carved well after his death in which he is likened to a jaguar god and to each of the hero twins of the sacred *Popol Vuh.*

Perhaps the most remarkable achievement of the ancient Mayan culture, which remained intact for close to a millennium before it declined rapidly in the ninth century A.D., was its intricate and detailed system of writing—particularly its mathematical numeration, astronomy, and calendar. Look at the dots and bars that make up the complex Mayan numeration system in the Dresden Codex (Figure 4-1). They were derived from the preliterate body-count system in which articles were tallied on fingers and toes, a time when quantitative information was conferred by gesture. A handful, for example, literally became a hand extended with the fingers held together. When some unknown tradesman first decided to encode such a quantity permanently, it became an elongated hand in profile and later an abstract dash or bar representing the number 5. Ones, like the imprints of the finger and toe tips that once tallied them, became dots. The completion of the full body count was represented by the open palm held against the jaw, or more simply by a closed fist. The latter was also depicted by its likeness in form to a stylized conch shell and in even more abstract form by an oval or lens-shaped symbol with a line through it. Whereas Babylonian numeration developed as a kind of economic shorthand, Mayan mathematics seems to have served a more lofty theocratic interest. In book form its means would follow the stars—true astronomy, while its ends served a New World form of astrology.

Epigraphers study writing—and those who initially pored over the Mayan inscriptions, both on monuments and in the handful of books that survived the Spanish Conquest, decades ago tended to view the Maya as a peaceful, theocratic people like so many wise old Greek philosophers clad in togas. Their most notable achievement lay in the realm of the abstract intellect; they were thought to have been selflessly absorbed, perhaps as no other culture on earth, with esoteric matters of astronomy and timekeeping, and to have developed a philosophy concerned with knowing the universe dispassionately and for its own sake.

As British archeologist Sir J. Eric Thompson eloquently expressed it: "[The] Mayan character, with its emphasis on moderation, discipline, cooperation, patience, and consideration for others, made possible outstanding achievement in the intellectual field. . . . So far as the general outlook on life is concerned, the great men of Athens would not have felt out of place in a gathering of Maya priests and rulers."[1] Thompson lived in the aftermath of the discoveries of the intellectual achievements of ancient Old World civilizations. His words echo the attitude of nineteenth-century enlightenment, which taught the white race to regard the Native American as a noble savage. Unlike ancient Greco-Roman texts, known for centuries, Mayan inscriptions had not been penetrated until the twentieth century, and they always have been cloaked in mystery. Since explorers in the time of Columbus first laid eyes on these people, many of whom still lived in and about the monumental ruins of their once exalted cities, the Maya have been depicted as exotic—somehow a cut above the savage races of the Americas that the early discoverers and missionaries wrote about. One could tell from the indigenous written books—filled with demonic-looking deities— that the Devil dwelled within them, as one early bishop of Yucatán stated, yet they computed via elaborate mathematics, recording periods of celestial bodies that ran to millions of years. And they developed a writing system consisting of nearly one thousand phonetic syllables— all in an environment that could not have seemed more hostile or less nurturing, at least to the European intruder. (Compare our twenty-six-letter alphabet, which generates far fewer sound variations than exist in the Mayan language.)

Today the study of ancient Mayan astronomy and calendars finds itself in much the same state as that of ancient Middle Eastern writing around the turn of the twentieth century. Perhaps we are still too close to the recent breakthroughs in document decoding to appreciate the degree to which the framework of our analysis is colored by the attitudes of the day. For example, the contemporary orientation to the ancient Mayan quest for astronomical knowledge is perceived in a context that is very different from Thompson's, one unabashedly rife with stressing mundane human need, moral but especially political. As art historians Linda Schele and Mary Miller put it:

> For the Maya, the world was a complex and awesome place, alive with sacred power. The power was part of the landscape, of the fabric of space and time. The king acted as a transformer through whom, in ritual acts, the unspeakable power of the supernatural passed into the lives of mortal men and their works. . . . The king ensured that the heavens would rotate in perpetuity through the rituals of sacrifice and bloodletting.[2]

Modern attempts to understand the Mayan written record have become more cross-disciplinary, our attitudes toward the people who wrote the documents so clearly changed. Today's New World historians regard blood, more than detached intellectual curiosity, as the mortar that welded together the scientific, social, and historical framework of Classical Mayan culture. Mayan astronomy was part of the promotion of the concept of the divine origin of kingship; it demonstrated the power, the regularity, the reliability of rulership, as I hinted in the discussion of the Palenque Triad. It served the needs of kings who regularly and publicly let blood by piercing their genitals with a stingray spine to demonstrate their bond with the gods of lineage. Should such a revelation tarnish our image of the Mayan astronomers? Our immediate answer would be no, yet we seem to judge discoveries in our own culture according to the way scientific information is put to use. For example, though the concepts and methods of nuclear physics are the same, we condemn or praise science for unleashing the power of the nucleus upon the world—whether that resource be employed to create weapons of destruction or alternative sources of energy.

Our problem with Mayan science, not unlike the problem with our own, seems to be to make sense of how scientific and social processes go together. How can we reconcile the precision and intricate detail that emerge from the Mayan astronomical record with what we still take to be primitive and superstitious ways in which astronomical knowledge was employed. This is an important issue, for it informs the way societies construct their understanding of the world. We are schooled to pay little attention to the cultural context that frames science—in this case the way ancient astronomy once fit within ancient culture. We will need to recast astronomy in its social framework to explore the cultural imaging technique that codified astral perceptions into language that instructed social behavior and brought order to the universe.

One idea in particular that we will need to reawaken in ourselves is that repeatable events can be perceived as parts of time cycles that fit neatly together. This cyclic concept of time, which advocates that the past contains the future, conflicts with our modern linear view of history, which tells us that all events are strung out on a "time line" of infinite extent, a one-way street with ever-changing scenery from the Big Bang to the present moment.

In some instances, repeatable celestial cycles were combined with those derived from social dictates. Take for example our system of Julian days, devised in sixteenth-century Europe and still in use by modern astronomers. It meshes the seven named days of the week, the period of return of a particular phase of the moon to a specific date in the seasonal year (nineteen years), and a fiscal period associated with tax collecting (fifteen years). More rare than pulling three bunches of cherries on a Las Vegas slot machine, the intersection of these three cycles on a single day will not happen again for 7980 years. Likewise, the Mayan Calendar Round of 18,980 days is the smallest whole multiple of a year, counted as 365 days, and a 260-day period, which in turn is composed of a pair of cycles of 20 and 13 days—like our monthly cycles of 7 named and 30 numbered days. This habit of building up smaller cycles to make bigger and bigger ones seems to have been shared by people all over the world. Only recently has it become submerged in our more linear way of structuring time.

As I explained in the last chapter, organizing time by seeking commensurable periodic relationships among celestial and terrestrial phenomena was quite typical among the cultures of ancient Greece, Babylonia, and the New World. What these highly structured societies had in common was the desire to construct a formal, dependable, and precise mechanism so that one event, say the setting position of the sun, could be used as a time check upon another, say the arrival of first crescent moon in the west after sunset (Figure 2-2)—or the somewhat less dependable appearance of the first rains or the seasonal return of a flight of birds. For example, listen to the way Hesiod in his *Works and Days* uses the sun and the simultaneous first appearance of stars and garden snails as time checks in the agricultural calendar:

> But when Zeus has brought to fulfillment the sixty-day period
> after the solstice of winter, a period of stormy weather,
> then the star Arkturos, leaving sacred Okeanos,
> brightly shines for the first time in evening's earliest
>   darkness.
> Next the swallow, the lamenting child of Pandion, appears,

91

coming into the sight of men when spring is beginning.
Keeping ahead of her, prune your vines, for this will be
  better.
But when the House Carrier goes up onto the plants from the
  earth,
fleeing the Pleiades, then no longer be hoeing your vines,
but be sharpening your sickles and rousing your slaves to
  their work.
Flee the seat in the shade and stay in bed until dawn
during the season of reaping when the sun is withering the
  flesh.
This is the time to be hastening and bringing your harvest
  home,
rising at daybreak, that you may have sufficient to live on.
For Dawn rightfully claims as her own a third of the day's
  *work.*³

The correlation of one sky event with another assured architects of
the calendar that should one phenomenon be obscured from view, a
point in time still could be reckoned through the observation of another. In a sense, one event would stand in as a substitute for the other.
If neither could be observed, then a third phenomenon might be assigned to keep nature's appointment.

This seems to be what is going on in many parts of the Mayan
Dresden Codex. Made of painted bark, this folded screen book may
have been among a sampling of artifacts sent back to King Charles V of
Spain by Cortés, though it did not come to light until 1739. Then it
was purchased in Vienna by a onetime emissary of the Polish state,
who was then the director of the Dresden library. There it lay for
nearly a full century until the famous explorer Alexander von Humboldt (described by one biographer as having been interested in just
about everything) got hold of it and published a few of its pages,
coincidentally the ones I will discuss in this chapter. That this portion
as well as others of Dresden's fifty-six leaves constituted an astronomical ephemeris (literally a daily listing)—a collection of data on the exact
whereabouts of celestial bodies—was demonstrated in great detail in
the 1890s by Ernst Förstemann, then the library's director. His was the
first attempt to take the astronomy in the Mayan codices seriously, at
least from a scientific point of view. Since then epigraphers have mined
a wealth of astronomical revelations from the diverse almanacs that
appear in these books.

But make no mistake about it. Even if astronomical observation and
mathematical computation went into them, books like the Dresden
Codex contained everyday knowledge for everyday people. However,
it was knowledge that could be transmitted only by the properly in-

doctrinated specialist or diviner who knew the sky and how to interpret the dialogue between people and planets. Priests once carried these books from town to town, telling their clients, noble and commoner alike, what action they ought to take to repay their debt to the celestial deities for conveying good things to them or to avoid their bringing pestilence.

Imagine these priests walking about the ancient Mayan city and its tributaries seeing royal clients and using their mathematical knowledge of the motion of the planets to minister to the astrological needs of the state. Says Bishop Diego de Landa, sixteenth-century Spanish chronicler of Yucatán:

> *They had a high priest whom they called Ah Kin Mai [literally a "day-keeper"]. He was very much respected . . . the lords made him presents and all the priests of the towns brought contributions to him, and his sons or his nearest relatives succeeded him in his office. In him was the key of their learning and it was to these matters that they dedicated themselves mostly; and they gave advice to the lords and replies to their questions. . . . They provided priests for the towns when they were needed, examining them in the sciences and ceremonies. . . . And they employed themselves in the duties of the temples and in teaching their sciences as well as in writing books about them.*
>
> *They taught the sons of the other priests and the second sons of the lords who brought them for this purpose from their infancy, if they saw that they had an inclination for this profession.*
>
> *The sciences which they taught were the computation of the years, months and days, the festivals and ceremonies, the fateful days and seasons, their methods of divination and their prophecies.[4]*

Even the lost books come alive in another chronicler's description, sounding just like the surviving Dresden Codex:

> *In these [books] they painted in colors the count of their years, the wars, epidemics, hurricanes, inundations, famines and other events . . . what prophecy there was about the said year and age; . . . it is all recorded in certain books of a quarter of a yard high and about five fingers broad, made of the bark of trees, folded from one side to the other like screens; each leaf of the thickness of a Mexican Real of eight. These are painted on both sides with a variety of figures and characters . . . which show not only the count of the said days, months and years, but also the ages and prophecies which their idols and images announced to them, or, to speak more accurately, the devil by means of the worship which they pay to him in the form of some stones.[5]*

We can only imagine the vast Mayan legacy that perished in the wake of the early fanatical Roman Catholic priests who followed the sword into Yucatán bent on converting the infidels. The destruction of pagan New World calendrical and astronomical documents was both widespread and thorough. One list alone tallies more than five thousand stone idols, and over two dozen hieroglyphic books. How many versions of the Venus table I am about to explore have yet eluded our eyes?

Recall that vivid imagery from Chapter 3 (Fig. 3-4e)—a Venusheaded deity hanging from the sky who appears at one end of an eclipse prediction table—a table based on the cycle of the moon, not the sun. The simple message to the people: Venus plunges earthward during the eclipse, bringing with him omens. But a more complex esoteric message, couched in the language of mathematics—a language not shared by all Mayan people—might read: The cycles of Venus and the moon are linked harmoniously; we the astronomers have proven it in our computations.

One of the most unusual expressions of the process of commensuration or cyclic buildup in Mayan calendrics surfaces again on pages that immediately precede the eclipse table—a set of events explicitly associated with Venus that are submerged in an ephemeris that times phenomena associated with the moon. At first sight, our minds might interpret these tabulations on pages 46–50 of the codex (Figure 4-1) as some magical manifestation of mysterious priestly knowledge, a secret attempt to garb one wanderer in the guise of another, the way we will find in Chapter 6 that Galileo recorded the viewed phases of Venus in a cryptic anagram. That would be true if we insisted upon seeing Venus marching in time to a solar beat, the way it appears in our current almanacs and planetary tables. These depict the rising and setting times for the planets according to the day and date of the year. Indeed, the year has been the principal long-term reference unit in our calendar since the Julian calendar reform at the height of the Roman Empire.*

Again I ask: What would a system of long-term time reckoning look like if the moon rather than the sun were the principal indicator? If the month rather than the year were the base interval, we could anticipate what a predictive table of astronomical events might show. For example, a listing of Venus's motions ought to reveal multiples of the month of the lunar phases (29.530589 days by modern reckoning) and the Venus synodic period (583.92 days). If the culture that devised it also had struggled with the imperfect fit between synodic month and seasonal year (365.2422 days), then we might further anticipate that a favored interval consisting of a whole number exactly divisible by all three key periods would appear.

---

* In the business world, economists also reckon the monetary cycle with its infamous "quarters" by solar time. Perhaps the conditions of extreme climate that occur in the temperate latitudes once played a role in propelling the seasonal solar year to its ultimate dominance. But we still make our (monthly) car payments according to a lunar schedule!

The number 2920 is an excellent candidate, because it is commensurate with all three cycles—the sun, moon, and Venus. It even neatly encapsulates the five unique forms of the track Venus makes in the twilight sky during a sequence of morning star and evening star periods.* In simple terms, five Venus rounds is about two days short of eight years and about four days shy of a whole number (99) of full moons. Although we cannot be sure whether Venus was integrated into it, the Greeks knew this 2920-day cycle very well; they called it the *Octaeteris* (connoting eightfold or octal), and it was one of the earliest combination cycles used by the Athenians to fit moon phases, by which they reckoned days and months, into seasons of the solar year. Their religion insisted upon carrying out the rites to the gods exactly on schedule, lest the member of the pantheon being addressed be displeased.

This habit of adapting one chronological cycle to another by recognizing two or more periods that fit together is shared by many cultures. For example, a Moslem astronomer of the Middle Ages tells the story that his people wanted to fix their month of pilgrimage at harvesttime permanently so that certain grains, fruits, and vegetables would be ready for the markets that were held to coincide with political discussions among the assembled tribes. So they decided to *intercalate*, literally to insert within the calendar an extra month on an occasional basis. To keep long-term accuracy, the bureaucracy in turn would require that this should be planned out well in advance, an effort that required an intimate knowledge of the lengths of various long periods that equaled whole numbers of months as well as years. Thus it was that astronomy came to the service of the Moslem state.

If the Mayan Venus calendar uses the five visible Venus periods or a 2920-day interval as part of a lunar rather than a solar time line, could we also expect that Mayan astronomers established components of the 584-day Venus cycle in time units based on the moon? Recall that the actual morning and evening star intervals that make up the Venus cycle pictured in Figure 2-4 average out to 263 days apiece, and that these were sandwiched in between unequal periods of absence, 8 and 50 days long on the average. If there were a lunar yardstick in the calendar, we might anticipate finding some sort of statement suggesting that each appearance is about nine months long. But how would the 50-day period, which is not close to any recognizable lunar or solar interval, or the unwieldy 8-day period be expressed? And what about all the other

---

* To state these relationships in quantitative terms:
Venus Rounds:
5×583.92 (5 Venus cycles) = 2920 days $-$ 0.40 days
Seasonal Years:
8×365.2422 (8 tropical years) = 2920 days $+$ 1.94 days
Moon Phase Cycles:
99×29.530589 (99 months of lunar phases) = 2920 days $+$ 3.53 days

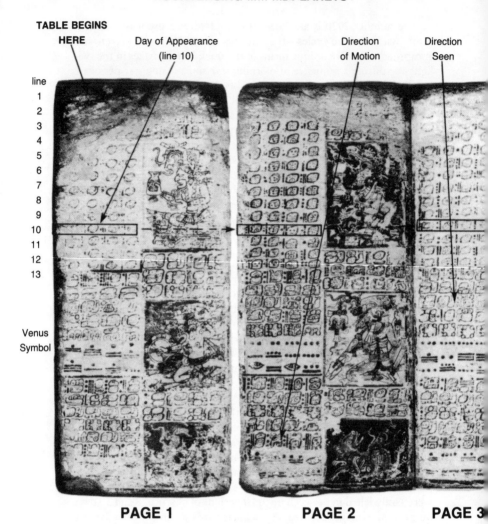

FIGURE 4-1a. VENUS IN A MAYA SACRED BOOK

The five-page Venus Table in the Dresden Codex showing pictures of the Venus god and his events, dates, intervals, directions, and resulting omens. Dotted arrow indicates the flow of time depicted across one of the thirteen lines of ritual day names assigned to Venus appearances and disappearances.

This cycle consists of two appearance and two disappearance events culminating in morning heliacal rise. As you move from left to right, for example, along the list of named "days of appearance" in blocked out

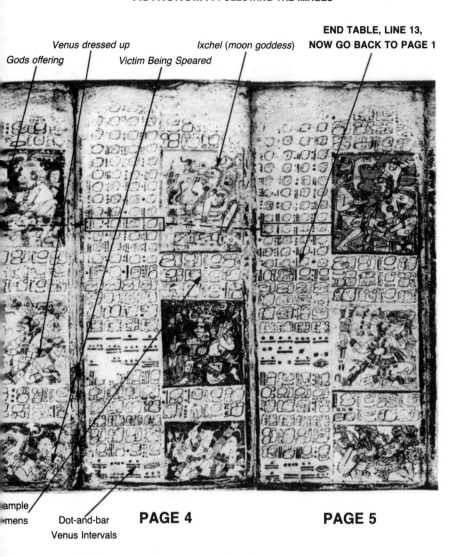

Gods offering

Venus dressed up

Victim Being Speared

Ixchel (*moon goddess*)

END TABLE, LINE 13,
NOW GO BACK TO PAGE 1

ample

mens

Dot-and-bar
Venus Intervals

PAGE 4

PAGE 5

horizontal line 10, you pick up the appropriate information about which way Venus was moving, and so on from the columns below each named day. An example is given in Figure 4-1b. Having concluded one 584-day cycle, you move on to the next page and follow the dotted arrow all the way across the five-page text. The omens for each cycle appear in hieroglyphic notation over the head of the Venus god, who is shown with shield and spear thrower. Pertaining mostly to warnings about health and livelihood, one set of such omens is translated on page 102. (Source: Graz: Akad Druck–Ü Verlag)

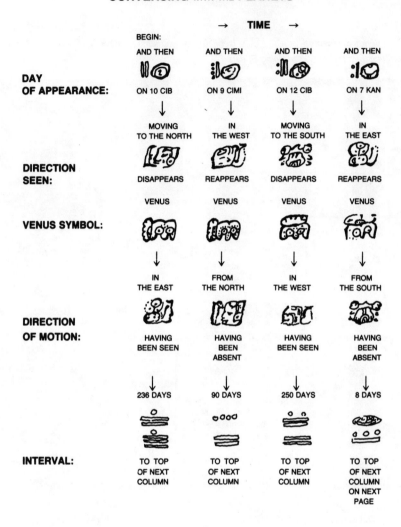

FIGURE 4-1b. Translation of a portion of the first page of the Venus table showing the divisions of time over a single 584-day cycle. The whole table incorporates 65 such cycles before it repeats itself.

periodicities that we might not associate with the moon or Venus in our way of thinking which could appear in such a calendar? Our game of temporal hide-and-seek in undecoded ancient documents could quickly degenerate into a random search.

Still, cycle-minded lunar calendar builders would have been on the lookout for rhythms tucked away in the natural environment that could, with some ingenuity, be manipulated to fit together, the way the Moslems wedged whole numbers of months into a specific span of years, or the way we alternately hone down or pad our seasonal years into sequences of 365, 365, 365, 366 to fit the seasons over the long run. Like the segments of the Venus cycle, no one of these intervals, taken by itself, is a precise approximation to the true length of the year, yet, averaged together in long temporal chains, they are exceedingly accurate indicators of the sun's real whereabouts.

A good look at the Venus Table in Figure 4-1 is enough to put off any serious student of modern science. What are those grotesque figures, long-snouted animallike deities doing there—those squatting, toothless old men with appendages sticking out of their foreheads? Why is Kukulcan, the Venus god, so demonic in appearance as he dons different outfits across the middle band? Who are those hideous half-human creatures at the top shaking what look like lidded jars—or those dead animals across the bottom span? What place does weird imagery like this have in an exalted table that purports to predict future natural events with great precision? We see no celestial metaphors, if that is what these are, on our modern weather maps or medical diagnostic charts. Even Mayan scholars have viewed Maya astronomy as astrology pure and simple. But it is our inherent naïveté about how others see and interpret the world that often can keep us from appreciating what their science could achieve.

As Otto Neugebauer suggests, it is only when we descend into the intricate detail (as the chapter epigraph puts it) of each of the five pages of the Dresden Venus Table that we begin to capture the mind of the astronomer at work, and to penetrate the Mayan intellect. After coursing through its pages, no reader will be left to doubt that the wise men who composed the Venus Table were conducting their dialogue with the sky not only in the poetic meter of myth but also in the precise and rigorous language of mathematics—the way astronomers still express what they see happening in the sky. What astonishes us is the way they mixed magic with mathematics.

Each of the five pages of the table,* executed in bright colors—eggshell blue and rust red—consists of a set of pictures arranged one on top of the other, cemented together by hieroglyphic symbols. To the left lie long columns of numbers and calendrical glyphs. At the bottom of each page are four two-digit numbers.* In Mayan base-twenty notation, the numbers increase from bottom to top. Enlarged in Figure 4-1b, the first of these numbers reads two bars and one dot on top of three bars and one dot, or sixteen 1s, plus eleven 20s, which equals 236.

---

* A dot equals 1 and a bar, 5. As I stated earlier, these are surely abstracted written forms of what once had been hand gestures, the dot being the point of the finger, the bar a completed handful of five. To count articles in the base-twenty or vigesimal system, let

Read in similar fashion, the second number is 90, the third is 250, and the last entry is 8. (Note the 0 curiously left in place over the three dots and bar.) Each of the five four-set blocks adds up to 584, leaving little doubt that the table deals with the Great Star—Venus Kukulcan, one end of the two-headed sky serpent. The division of the period into four parts emphasizes the quadripartite nature of the Venus cycle as the Maya conceived it: the interval as morning star (which they put at 236 days), a lengthy period of disappearance in the glare of the sun (90 days), the interval as evening star (250 days), and a second relatively brief disappearance (8 days). The scheme seems to be some sort of corruption of the intervals actually observed: 263-50-263-8, which I mentioned in Chapter 2.

Longterm time flows from left to right across the five-page table, and pictures always follow the eight-day disappearance of Venus before his predawn return to the sky. There are five Kukulcans all with different outfits, one for each of the *five* manifestations of Venus; thus there are five ways the great omen bearer can fling his ominous spears, bright darts of light, at victims who lie impaled below him in each panel. For example, the one at the center of the third page is part human, part jaguar, with blackened eyes and nose, clad in the skin of a jaguar. We can identify him with an Aztec deity named Yoaltecuhtli, or guardian of the night sun, an apt role for Venus. On the fourth page, the central spear thrower is a blindfolded white god, probably a form of Tezcatlipoca, god of cold weather and another form of Venus as god of the dawn. The writhing targets of daggerlike Venus omens, represented in the pictures at the bottom of each of the five pages, seem to depict human and animal forms of nature gods in postures affected by the ominous radiance of Venus. For example, the youthful god of maize is featured on the third page—his necklace of kernels and corn-tasseled headdress are easily visible—and on the fourth an amphibian-headed creature squats. If the latter is a tortoise, then he may be a bringer of rain for, as one contemporary Mayan legend has it, when he weeps for the plight of the farmers, his tears cause the rain to fall. (To reciprocate, the serious Mayan nature worshiper is still required to warn all turtles to find safety at the time he burns off the remnants of last year's cornfields in anticipation of the rainy season.) Finally, the five pictures across the top probably represent offerings made to the Venus deity. Except for the frame on the

the lowest order be 1s, the next highest 20s, the next 20 × 20 or 400s, the next 20 × 20 × 20 or 8000s, and so on. When counting time, however, the Maya altered the third place to 18 × 20 or 360, specifically to approximate the number of days of the year. For example, the number 365 in Maya as a count of human beings would be $5 \times 1 + 6 \times 20 + 3 \times 400 = 1325$ people; as a count of time, however, it would be $5 \times 1 + 6 \times 20 + 3 \times 360 = 1205$ days.

fifth Venus page, which depicts two figures talking over the situation, every other illustration shows a seated figure offering up incense, plants, and so on.

There is more quintessential rhythm making going on in the scheme for naming the actual days when the Venus events take place. The top section of each page—the first thirteen horizontal lines—consists of an array of hieroglyphic representations of the twenty named days of the Mayan "week," each preceded by one of the thirteen numbers. Taken together, they make up the sacred Mayan 260-day round. We can think of these entries in the same way we might think of labeling an astronomic event by the day of the week and month it occurred, for instance, "a full moon happened on Friday the thirteenth." The arrangement of these named days and numbers, each one of which goes with the disappearance interval at the bottom of its vertical column, indicates that the user of the table would have entered Venus time on the first horizontal line at the top of the first page, his eye passing all the way across each page through the end of the first line of the fifth page; then he would proceed to the second line of the first page, and so on until he had completed the thirteenth line on the fifth page. (Imagine scanning across the first line of several pages of the text you are now reading before beginning a second line!) In reality, the completed table really comprises 13×2920 or 37,960 days, about 104 years (a Great Cycle, so called because it is a whole number of Venus cycles, 260-day ritual cycles and 365-day year cycles). At the end of this Great Cycle, the table can be reentered in step.

What do the other symbols mean? The most important ones that lie between the thirteen horizontal lines of day names and numbers and the interval numbers include a set of glyphs that indicate the direction in which Venus is moving during a given interval of its orbit, and another set that tells where it will be seen. Threaded across the middle of each page are forms of hieroglyphic symbols of the planet Venus— some are enlarged in Figure 3-4b–d. One looks like a pair of eyes gazing out at us from the numbers and pictures in the table. For example, notice how the glyph is suspended from the headdress of the Venus deity on the second page immediately to the right of the Venus symbols tabulated on that page. (I have labeled each of these segments to aid readers in following the table.)

We find the Venus glyph elsewhere in Mayan writing, sculpture, even painting. For example, another set of Venus-related pictures appears in sky scenes above a group of mural paintings inside a temple at the ruins of Bonampak, not too far from Palenque. There, Venus hieroglyphs ride on the backs of weird animal and human creatures who represent signs of the constellations of a Mayan zodiac that chart the course of Venus along the stellar roadway. The Venus imagery at Bonampak is identified with dates given in the associated inscriptions on monuments at the site. These dates match actual first morning appearances as seen from that place in the eighth century A.D., when the paintings were made. But what does Venus bring in its wake? Like the omens written between the pictures of the Dresden Codex, the answer lies in the subject matter of the murals, but we do not know enough about the written and pictorial record from ancient Mexico to understand why, of all planets, Venus was chosen to be the one associated with the conduct of war ritual and the initiation of battles.

Suppose we make a literal translation of a short segment of the table, beginning with the set of four day names in the 260-day round in the tenth line of the first page. Details are blocked out in Figure 4-1b. For simplicity, I have omitted much of the extraneous information, such as specific colors the Mayans associated with the directions, month names, and so on. For each day name (one of twenty) and associated with day number (one of thirteen), the table announces a key position in the Venus cycle, telling which way Venus was moving, where it appeared or disappeared, and how long it was absent. The reading at the first day name on horizontal line 10 picks up where line 9 has just ended, on the fifth page of the table. It continues: *And then on the day named 10 Cib, moving to the north* (six lines below and in the same vertical column), *Venus* (two lines farther below in the same column) *disappears in the east* (seven lines farther below in the same column) *having been seen for 236 days* (two lines farther below). The next stage of the cycle happens on 9 Cimi (line 10, second vertical column), and the text continues in the same manner: *And then on 9 Cimi, in the west, Venus reappears, from the north, having been absent 90 days.* And the next: *And then on 12 Cib* (line 10, third column) *moving to the south, Venus disappears in the west, having been seen 250 days. And then on 7 Kan, in the east, Venus reappears from the south, having been absent 8 days.*

At this stage of the cycle bright Venus reappears the sky as morning star (at the position indicated in Figure 2-4), the shield-bearing Venus god in one of his five big-picture manifestations. The event is represented by a dozen blocks of glyphs arranged over his effigy in the middle of the page. These encode the omens for this particular set of Venus motions. One set, for example, reads:

1 HE IS SEEN	2 IN THE EAST	7 WOE TO THE MOON	8 WOE TO MAN
3 GOD (SHOWN AT MIDDLE)	4 VENUS	9 THE DISEASE	10 OF THE SECOND MAIZE CROP
5 GOD (AT BOTTOM)	6 IS HIS VICTIM	11 WOE TO THE MAIZE GOD	12 WOE TO NIGHT

Basically the omens seem to inform the user that once Venus reappears in the east the god of maize or night or certain diseases will be "his victim"; in other words, bad crops, pestilence, or some other misfortune will transpire. Although the translations of all the omen hieroglyphs are woefully incomplete, it seems apparent that to the Mayan intellect the ills of the world are aligned with sets of deities who differ on various occasions. It does appear odd that all the Venus omens are malevolent, especially when we realize that most divining systems consist of a better balance of good- and bad-luck days.

The Venus omen text then moves ahead to the block of four Venus dates on the next full page, the second page of the table, continuing with the day name on line 10: *And then on 9 Ahau*, thus encountering each set of five pictures as it runs horizontally to the conclusion of the full run of five pages along line 10. I have indicated this run of the table by the dotted arrows in Figure 4-1a. Next, the priest's eyes would pass back to the first page for the next runs across lines 11, 12, and 13. Approaching the sixty-fifth 584-day cycle (5 pages × 584 days per page × 13 lines), he would make an appropriate correction indicated on the user's page (not shown) that precedes the table. This would restore any temporal slack between the average Venus period of 584 days given in the table and the true Venus period as seen in the sky, which, you will recall, is a fraction of a day shorter than 584. Only then could the diviner recycle back to line 1 of the first page (effaced by water marks in Figure 4-1a) and thus mete out another century's worth of Venus predicting—to an accuracy of one day in five centuries!

In the lower section of each page of the table, there are matching sets of "base dates" in the 365-day calendar that can be selected and meshed with those from the 260-day calendar above to give the complete Calendar Round date of a Venus appearance or disappearance. It is on these dates that we are offered a clue to the strange rhythmic lunar beat that underlies the structure of the table. The three sets of base dates in the 365-day calendar are separated by time intervals that are divisible almost exactly by lunar eclipse cycles. Evidently, the cycle-minded Mayan astronomers had discovered one of their gods' top secrets—how eclipses and Venus's appearances fit together in a harmonious, predictable way. Here was a pattern of celestial order unknown to their predecessors, even perhaps to their Old World counterparts.

What do I mean by an eclipse cycle? Two periods are involved. First, there must be a new moon (for a solar eclipse) or a full moon (for a lunar eclipse). Second, the moon must be situated at a crossing point, or node, of the earth's orbit about the sun, so that all three line up in space. Curiously enough, this latter interval still carries the name *draconitic month*, after the reference to the mythical dragon who was alleged to have devoured the sun or moon during eclipse. Of course, whether or not we think of this cycle in terms of orbits crossing, its recognition would have been seminal for eclipse prediction.

Anyone who has ever seen an eclipse, especially a total eclipse of the sun, cannot fail to be impressed. And the Eclipse Table that immediately follows the Venus Table in the Dresden Codex shows that, by counting and grouping full moons in sets of fives and sixes, Mayan astronomers could predict with regularity when these spectacular phenomena would occur.

Juxtaposed, the Venus and Eclipse tables can also be related numerically. The length of the whole Venus Table is just three times the length of the Eclipse Table plus eight cycles of 260 days, a choice of cycles that also emerges as a deliberate aspect of computational astronomy.

Associating eclipse prediction with Venus appearances would be a logical outcome of the discovery of commensurate lunar and Venusian cycles. But what does it mean on a practical basis to chart the course of wandering Venus with a lunar yardstick? The *octaeteris*, the duration of one line taken all the way across the Venus Table, equivalent to eight years, helps us understand how the Maya did it. On a practical basis, Venus events would fall back by about four days every ninety-nine full moons (see note on page 95). After thirty-two years, for example, a first morning rise will have shifted halfway through the lunar cycle, and, if we double this interval, it will be shifted through one full cycle.

Imagine then that a priest visiting your village to cast omens by the appearance of things in the sky actually had to work with such a calendar to make predictions and set celebrations. How would he do it? He might be tempted to intercalate extra days to fill in the shortfall. But such a procedure would not really be necessary if he knew how to switch from canonical time (time by the calendar document) to real time (based on what he saw in the sky). If he knew his position in the thirty-two-year cycle, he ought to be able to figure in his head how many days to cut off in a Venus cycle to predict what the condition of the planet would be.

At the bottom line of each of the five pages of the Venus Table we appreciate even further the lengths to which Mayan astronomers were willing to go to distort short-term Venus time from reality so that, in the long run, its cycles would fit perfectly with the lunar measure they believed lay at the heart of their timekeeping system. Once again, "moon numbers" lurk in the four Venus intervals that make up each 584-day cycle. Remember that three of the four subintervals are only very rough approximations of the actual intervals of appearance and disappearance of the planet. They must have been altered deliberately to correspond to lunar and possibly other ritual dictates not yet known to us. These intervals also are approximately equivalent to the multiples of the moon phase period. Thus, the morning star interval of 236 days is just a quarter day short of eight full moons, and the 90-day disappearance period lies close to three full moons, while the 250 days the astronomers officially assigned to the evening star period is about a day and a quarter less than eight and a half lunar months. (The sole exception is the eight-day mean interval of disappearance before morning first rise.)*

What do these corrupted or aberrant intervals mean in practice? And what did the astronomers have in mind when they used them? Suppose the moon were at a particular phase when Venus made its last appearance as morning star; then, after the first interval given in the table (236 days) the astronomer would know that the moon would necessarily be in the same phase and Venus would make a first morning appearance. After the next interval (90 days) the moon would still be in the same phase and Venus would make its first evening star appearance. Following the third interval (250 days), the astronomer would know that just as Venus was about to make its last evening disappearance, the moon would be in the opposite phase; in other words, if it had been first quarter it would now have changed to last quarter. Finally, irrespective of the lunar phase, the last interval would count the eight days on the average to the reappearance of Venus in the morning sky.

If the Mayan Venus-Moon calendar seems a bit mysterious, think for a moment about the constraints in our own system: We celebrate most holidays on a Monday in order to lengthen our weekend. Thanksgiving is always the fourth Thursday of November, yet the Fourth of July remains the fourth. Like the architects of our civic calendar, Mayan timekeepers had limitations fixed upon them by the oft-conflicting demands of religious belief and the desire to follow nature's perfect metronome. We can understand their plight if we imagine that, instead of scheduling a fireworks display, we needed to time the celebration of American independence with a more naturalistic celestial spectacle, such as a meteor shower. Clearly we would need to set up new rules and modify old symbols—perhaps counting, according to some fixed prescription, forward to the nearest expected or backward to the most recently observed meteor shower.

The moon is also present in the Venus Table in pictorial form. A lunar goddess is portrayed at the top of the fourth page of the table. Ixchel, whom the Maya called "Our Mother" or "Lady," a Mayan goddess who doubtless merged identities with the Virgin Mary after Spanish contact, had many guises.† She was also a patroness of childbirth and medicine, not to mention earth, trees, and the growth of plants. First wooed by the sun, whom she wed, she later had an affair with the sun's brother Venus. She still quarrels with her husband over

---

* (First column) 236 days = 8.0 lunar synodic months − 0.24 days
  Morning Star Apparition Interval
  (Second column) 90 days = 3.0 lunar synodic months − 1.41 days
  Disappearance Interval After Morning Star
  (Third column) 250 days = 8.5 lunar synodic months − 1.25 days
  Evening Star Apparition Interval
† In Conquest times the moon goddess, wife of the sun god–creator, came to be associated with the Virgin Mary, specifically with the divine umbilicus that links people to the one true god in the adopted Roman Catholic faith.

her infidelity every time there is an eclipse. On the fourth page of the Venus table, she seems to be seated on a celestial throne, emerging from a large moon sign, and she holds a conch-shell cup to symbolize her birth. A name glyph above identifies her. Whether she appears in the Venus Table to portend an eclipse we cannot say. That she is in one of the upper pictures may signify her role as mediator—perhaps placed there to placate the Venus god with an offering.

We know next to nothing about how the ancient Maya may have reacted to Ixchel's face being blackened. After all, the Venus and Eclipse tables were used three centuries before Columbus, and the information that went into making them up may have dated to a millennium earlier; however, a passage about lunar eclipses and childbirth from much later Aztec astronomical lore, as related by a Spanish priest, may offer a clue:

> When the moon was eclipsed, [its] face grew dark and sooty; blackness and darkness spread. When this came to pass, women with child feared evil; they thought it portentous; they were terrified [lest], perchance, their [unborn] children might be changed into mice; each of their children might turn into a mouse.
>
> And because they feared evil, in order to protect themselves, in order that this might not befall [them], they placed obsidian in their mouths or in their bosoms, because with this their children would not be born with mouths eaten away—lipless, or they would not be born with noses eaten away or broken off; or with twisted mouths or lips; or cross-eyed, squint-eyed, or with shrunken eyes; nor would they be born monstrous or imperfect.[6]

Is it possible to assign real dates in our framework of time, such as October 18, 901, to the dates given in Mayan notation in the table? If the answer is yes, then we can explore exactly what their astronomers were predicting and how accurate they were. Recently, linguist Floyd Lounsbury discovered a clue in one of the Long Count dates on the user's page. It registers the number of days lapsed since August 12, 3113 B.C., when the Mayan gods were said to have created the present version of our universe. If this date, presumably derived from ancient observations of Venus, is used to enter the table and if the diviner then passes through three full century-long runs of the table, the next reentry into the beginning of the table falls on 10.5.6.7.0 in Mayan time, or November 2, 934, in our calendar. This date fell exactly on the first morning appearance of Venus and on a day name in the 260-day count considered to be lucky for a Venus appearance. Consequently, Lounsbury argues that this was the likely date when this particular form of the Venus Table was installed.

By the tenth century, astronomers had already observed Venus's wanderings closely for hundreds of cycles. They must have become aware of the visible shortfall between the real Venus period of 583.92

days and the tabular version of 584 days they had been using. As it lost 0.08 days against every tabulated Venus period, the great star would have appeared about a day earlier every twelve cycles (about 20 years), or 5 days over the 104-year "Great Cycle" length of the table. As the user's page prescribes, the Mayan astronomers remedied the discrepancy either by knocking off 4 days or by assessing a double correction of 8 days every time they reached the end of one or two Great Cycles in the table, respectively. The choice of multiples of 4 would even preserve the correct lucky name day for all future starting events in the table and furthermore render the table inaccurate to just a few days over several centuries. The whole scheme would be analogous to tacking our leap year onto the calendar at the appropriate year's end instead of at the end of February, which, incidentally, once was the last month of the Roman year.

Since the Dresden document itself dates from about two centuries later than the 10.5.6.7.0 date, we can imagine the problem the Mayan sky watchers, long encumbered by ritualistic constraints, had been confronting for so many years as they attempted to devise updated versions. Above all, they needed a reliable methodology for anticipating the all-important reappearance of Venus in the eastern predawn sky after its brief absence from view, for only then could the omens make sense. The problem they faced is not so different from the one confronting Western timekeepers from the Roman Empire to the Renaissance that ultimately resulted in the Julian and Gregorian calendar reforms. In each instance, a date shift—in our case restoring the date of the spring equinox to the appropriate calendar date—was accompanied by the design of a set of rules (the famous leap year regulations) to reduce the drift between real and canonized time. The only difference is that Julius Caesar and Pope Gregory were addressing the conflict between solar and seasonal time whereas the Mayan cast of celestial characters played on a different stage. And, for Gregory at least, the motivation was to determine when to celebrate Easter Sunday.

As I have already hinted, all the embedded resonances between Venus and the moon in the Mayan record arouse suspicion that there may have been a way Mayan astronomers could actually have kept track of lunar eclipse events within the reference frame of their lunar-based Venus calendar—in other words, linked real Venus appearances with observable eclipses. But what was the mechanism and how did it function?

We can test various possibilities since we now believe we know when the table was installed by the priests, and we can back-calculate when Venus appeared and where and when eclipses of the sun and moon took place. To do so, we can specifically consult modern astronomical tables of eclipses that actually took place in the sky over ancient Yucatán and in addition immediately preceded and followed the corrected base dates of Venus's first appearances given in the table. When we do, we dis-

cover that a lunar eclipse was visible on the lunar phase cycle immediately preceding each of the first six starting base dates in the table.*
But, for the last two reentries into the table, which fell in the far distant
Mayan future (in the thirteenth and fourteenth centuries), no association between a lunar eclipse and a first appearance of Venus occurs—
and for a very good reason. By this time the Mayan culture as we know
it from the Classical Period, with its great works of sculpture and
architecture, had declined, and with it the interest in precise astronomy
became greatly diminished. This development probably accounts for
the failure of the table to anticipate Venus events by the time of the
great Mayan collapse.

So Mayan astronomers had discovered a sacred hidden parallel between two of their sky deities—the moon and Venus. By grouping
cycles of full moons, they could mathematically assure that every
eclipse would be scheduled to occur at a first rise of the great star. Our
deep descent into some of the rigorous detail in the Mayan Venus
document demonstrates just how diverse nature watchers can be. It
shows that it is possible for another culture to devise rather complex
and elegant principles that, though unfamiliar to us, are capable of
making nature work for them.

To our knowledge no chronicler of the New World ever recorded a
calendar priest in action; therefore, we can never know what a Venus
omen–bearing session would have looked like, but anthropologists like
Barbara Tedlock, who have worked with the living descendants of the
ancient Maya, imagine the contemporary process of divining to be not
so different from, if somewhat less elaborate than, what went on in the
heyday of the Mayan Classical Period.

Today's daykeeper sits at his table, arrayed with lighted candles,
bowls of incense, pieces of woven cloth, and other amulets pulled from
his divining bag. He prays to the now Christian version of his ancient
god to ask permission to divine. With no codex to guide him and with
little precise knowledge of the Venusian orbit surviving to make a
difference in his predictions, he arranges piles of seeds and crystals in
separate groups, consisting of equal numbers of elements. Then he
"uses his blood," described as a certain feeling in his upper and lower
arms, to tell when something is about to happen. He speaks of bor-
rowing the power of certain days to make his predictions. His performance is based on reading the seeds and crystals, which represent the

* To give an example of how this worked, a Venus event that marks an eclipse would
always take place after the beginning but before the termination of the next lunar cycle,
i.e., between fifteen and forty-five days after the visible eclipsed full moon (Aveni, *The
Sky in Mayan Literature*). In the vicinity of the first base date (February 6, 623), a lunar
eclipse was visible in Yucatán on January 22. Sixteen days later, or one day into the
cycle of the next moon, Venus disappeared. Typical for a February disappearance, the
planet was absent from view for about three days, returning to view on February 20.
The reappearance date anticipated by the table was February 6, or fourteen days
before the event.

names and numbers in the 260-day count still in use in rural Guate-mala. Next he answers a series of questions posed by the client: Will my sister come soon? Will the marriage of our daughter be favorable? Is my dead father truly at rest? Given the marvelous detail on the Dresden Venus pages, we can only imagine what a profound role the planetary pathways once must have played in this intimate dialogue mediated through the queries of the keeper of the days.

The Venus Table in the Dresden Codex is a subtle and splendid remnant of ancient Mayan mathematical and scientific prowess—a hall-mark from the height of Mayan intellectual achievement. When we look at it as we have in this chapter, the Maya seem closer to us. It is as if they share our scientific, quantitative way of comprehending the world. It is only when we pause to read the omens attached to the numbers in the table (like those I quoted on page 102) that we become sobered about our differences.

Mayan blank verse has the same syncopated and repetitious quality as the Hymn to Inanna in Chapter 3. The numbers tell us that the moon and the morning star still consort together, but the words endow the table with that metaphysical air modern readers find so difficult to reconcile with any brilliantly precise corrective clockwork. The thought of obsessive Mayan fatalists cringing helplessly beneath a sky filled with incipient doom replaces our carefully honed image of confident scientists in full control of nature, their codices in hand. We are per-haps still as baffled by the paradox of mathematics and superstition coming together as Bishop Landa who, when one of the Mayan codices was first brought to him, was heard to utter that all these works con-taining lies of the Devil must be burned immediately. Yet that same holy man, who destroyed thousands of Mayan books in a churchyard bonfire in the middle of the sixteenth century, spent the rest of his life describing the system of numeration and writing that had catapulted the ancient Maya to great scientific achievements.

We cannot resolve the paradox without understanding the people and the culture who created it for us. We have seen what Mayan society was like in its prime, when it spawned the original (now lost) version of the Dresden Codex. For one thing it was highly stratified, with peasant farmers living in the rural areas about a huge monumental ceremonial precinct. Closer in resided the merchants and artisans, and at the center itself lay great palaces inhabited by nobles and royalty, most of them, like the Habsburgs and the Bourbons, blood related. The astronomers were the Mozarts and Haydns, civil servants who were highly specialized and often attached to particular job descriptions as dictated by the state religion.

Eighth-century Copán, in the southeast Mayan region (today Hon-duras), spawned a Venus cult of its own. All over its dedicatory ar-chitecture we find Venus symbols and Venus events, many of them related to dates and depictions in the Dresden Codex. In no city in the world was history as written by the ruling elite so tightly intertwined with planetary astronomy. For example, a monumental text on one of Copán's altars, located in front of a sacred temple, has a first predawn appearance of Venus tied to the date of inauguration of a king whose name was New-Sun-at-Horizon, or Yax Pac in Mayan phonetic writing. Yax Pac is especially interesting because he appears to have ruled at a time of proliferation of large ceremonial centers that are believed to have signaled the institution of the *ahau* principle. This novel idea held that the king himself, now endowed with a new title, was conceived supernaturally and that he derived his power directly from nature— specifically from sky deities. Many Mayan archeologists and epigraphers find it plausible that an ideological transformation—one that carried the Mayan city-states of this period from an egalitarian to a more strictly hierarchical social system—was taking place. It is likely that this phase of Mayan civilization also produced the elaborate astronomical tables in the codices that we just discussed.

Another date referring to Yax Pac's rule is accompanied by a Venus glyph that says, "It shined, the Great Star—Lord Ahau." The inscription is written in hieroglyphs on the eastern doorway of his temple. Modern calculations show that the date given coincides with the actual time Venus would have been sighted from the doorway.

Completed in the second half of the eighth century, Yax Pac's temple is a virtual three-dimensional map of what the Maya imagined the cosmos to look like. That must have awed the Mayan peasants who came there to worship their rulers. The south side carries symbols of the underworld, the north those of the afterworld, with the middle world of today, containing the king's throne, sandwiched in between. Philosophically, the Mayan underworld was an exotic plane of existence through which rulers and diviners could pass only while in a trance. It was a real world with real, if supernatural, inhabitants, and every night when the sun set the underworld flipped over and became the sky dome. Little do modern tourists realize that when they pass through the north portal of Yax Pac's temple they are entering the belly of the Mayan underworld.

Historically, the Venus appearance referred to in the inscription coincided not only with Yax Pac's own accession to the throne but also with a similar event in the life of his father, whose name translates phonetically as Smoke Monkey. This was a first evening rather than a first morning rise, but because the two Venus risings would have occurred on the same day of the seasonal year, it seems likely that Yax Pac (or his clever court astronomer) had devised an elegant celestial spacetime symmetry to reinforce his claim of genealogical descent from his father. He had publicly declared that the heavens had behaved the same way for both of them. His future was written in the stars.

Yax Pac would tie most of the historical statements in the Copán inscriptions that pertained to his reign to the movements of the planet

Venus, particularly its occurrence as evening star. His accession, the conduct of war, the dedication of stelae and works of monumental architecture—practically everything he accomplished—are recorded under the sign of his patron planet. For example, the dedication date on the Great Ball Court at Copán, equivalent to January 6, 738, in the Christian era, marked the point of highest elevation of Venus as evening star in the western sky after sunset. Like a true astrologer, Yax Pac manipulated history and guided his behavior and actions by what he saw in the heavens, and he did so with profound consistency.

For the young Yax Pac, the cycle of Venus was more than a convenient, recognizable period in the state calendar, a device by which to anticipate drought or schedule coronation ceremonies. Venus was part of his reason for being—a celestial deity whose cyclic behavior mirrored in miniature the even longer cycles of creation that made up Mayan time. Here was another imaginative way of expressing the philosophy of repetition and recursiveness that weaves together all of Mayan cosmology. For the true believer, the planet Venus was a vertebra in the backbone of the structure of Mayan history.

But would Mayan astronomers have made mythological associations between Venus and dynastic history or seasonal events such as rain and maize planting if they did not have some reason to believe Venusian aspects bore a direct relation to real-life experiences? Here is where observational astronomy enters the picture. As I showed in Chapter 2, Venus's intervals of disappearance do indeed vary in a way that depends very critically upon the month of the year and the place on the horizon where the planet is sighted.

Yax Pac's grandfather, 18 Rabbit, constructed in the west wall of his Temple of Venus a long slotted window to serve as an aid in marking the time Venus would reappear in the sky following its last appearance in the west precisely during the rainy season. When Venus was sighted in the window, we can imagine the public assembled to view their lord, seated on his throne in front of the temple studded with Venus symbols—the same ones we found in the Dresden Codex—and draped with the effigy of the two-headed sky serpent. The awestruck people witnessed their king draw his celestial sustenance from Venus as it passed over the western end of the building during the performance of the royal bloodletting rite that bound him to the gods and permitted him to commune with them by penetrating the underworld plane.

In fact, we can even determine from measurements made on the building that Venus passed through the window while it was on its way to or returning from one of its extreme setting positions along the horizon, which are reached at eight-year intervals, but always during April and May, when the anticipated rains usually descended to nurture the crops.

The Venus horizon extremes occurred at the same time of the year as the traditional period associated with the onset of the rainy season— and 18 Rabbit's astronomers knew it. The Venus-rain equation, then, has a factual basis in both astronomy and the planting cycle. It helps explain both the orientation and the complex symbolism that appears sculpted in the frieze over the doorway of the temple. We can understand why the S-shaped water signs adorn the body of the Venus sky serpent and why a frog-headed mask (the frog is a water symbol) faces the doorway on each side. The so-called Cauac glyphs surrounding them also pertain to rain and storms. And the pendulous-nosed Chaac, Mayan god of rain, appears on the building in the form of masks set into the corners of the walls, while the head and torso representations of the young maize god, nurtured by the spring rains, are ranged along the upper part of the west wall.

All these symbols tie agricultural fertility and water to Venus in a sensible way. Recall that in the Babylonian myth of Inanna's descent it was the water god who helped her reemerge from the bowels of the earth. So, too, the water-Venus connection was heralded in ancient Mayan mythology, and in both cultures Venus became the mediator between the sun and water.

Although our modern computers can regenerate with relative certainty the sky events that unfolded above the ceremonial center of Copán that the Maya may have recognized, we can only speculate about how the people there actually behaved when Venus appeared over a special temple that was dedicated to a planetary god. Perhaps 18 Rabbit actually sat on his throne over his glyphic text, elevated and enframed by the two-headed sky dragon, his youthful rulership celebrated and symbolized by being likened to the first stage of the growth of the maize plant. As he became enthroned he sprouted and grew like new maize. To onlookers below he would have seemed to be entering the world of the living through the mouth of the divine sky serpent as his patron planet sprang into view over the deceased sun in the west.

The ceremony, like the careful timing in the calendar of which it was a part, had both public and private aspects. At just the right time, the public eye would confront two kinds of visual imagery, two identical notions of reality. Their king sits on the throne, embodiment of Venus in the sky and the sun just below the horizon. But the citizens of Copán witness the visual phenomenon of the sun attached to Venus as the two twist about in the western sky. In mythic time the Venus and sun symbols remain frozen in the stucco imagery carved at opposite ends of the two-headed serpent over the doorway.

Only the elite priests would have access to the knowledge gleaned from appearances of Venus framed by the window of the Venus temple, enabling their calendrical prediction as well as the celebration and reaffirmation of the *ahau* myth to proceed. Thus, the temple window became an observatory, a tool for acquiring information from the gods, the Maya's planetary guideposts, who revealed themselves to be the true source of power and authority, the legitimizing agents in a belief system embedded in both the public and the private sectors of Mayan society.

At Copán, the invention of the Temple of Venus was a stroke of genius, part of an elaborate system that related the center of worship to the natural and social periphery of the city, the sun to Venus, the king to deity—simultaneously and for all to witness. The ceremonies attending the astronomical phenomena pivoted about the king. However we might imagine them, they surely would have impressed anyone standing in the East Court of the Copán Acropolis over a thousand years ago.*

The use of ceremonial architecture as a pipeline to convey celestial messages to a throng in the midst of celebrating a ritual seems to have developed into a fad in the Mayan world at this time. There are numerous examples of natural events that likely were staged in this way, with inaugurations, celebrations of victories in battle, great royal turning points being commemorated on days when some important celestial event occurred. Recall the celestial spectacle at Palenque that I discussed in Chapter 3. Such elaborate theatrical stagings reveal Mayan beliefs about the essence of heavenly power in a direct and forceful way. The goal of specialized Mayan ceremonial architecture seems to have been to instill in the viewer-participant the same sort of passion that might well up in the breast of the medieval Christian pilgrim who saw the sun shine through the stained-glass windows of the cathedrals of Chartres, Reims, or Cologne for the first time or the patriot today who sees flags flying at a formal military occasion. For the ancient Maya, planetary phenomena carried powerful messages.

ISHTAR AND SIN

We find an almost identical wedding of science and myth in the ancient Middle East. In ancient Sumeria, the moon (Sin), a bearded male; the sun (Shamash), a female and child of Sin; and Venus (Ishtar), also a female, constituted one of the great celestial triads, as I mentioned in Chapter 3. Surprisingly, Sin was the most powerful. He traveled about the sky in his crescent-shaped boat while Shamash ruled the seasons and Ishtar mediated between them.

* Recently the University of Texas's Linda Schele has followed the Copán tradition of targeting Venus dates as a way of interweaving history and astrology all the way back to the foundation of Yax Pac's dynasty in the early fourth century. At least the dates on foundation monuments seem to be well correlated with key points in the observed position of Venus, especially with the highest point of appearance in the sky (shown in Figure 2-4). And many of the subsequent dates and events are identical to those in the Venus Table in the Dresden Codex.

But how can the moon dominate the sun if the latter's rays are more prominent? Perhaps the answer lies not in who regulates the power of night and day but rather in who really controls the longer cycle of repeating months. Having explored the lunar measure of time in a Mayan hieroglyphic picture book, I now turn to a Middle Eastern text that also meticulously and scientifically measures out the movement of Venus in moonbeats.

Closeted away in the British Museum is a series of seventy special Babylonian cuneiform tablets, so named because of the wedge-shaped elements that make up their message. These shorthand notations, hammered in moist clay and baked in an oven to preserve their form, have an unusual history. Before 3000 B.C. Babylonian merchants used clay tokens of various geometrical shapes to signify quantities of diverse articles of trade. These were transported in sealed clay envelopes embossed with symbols of the contents. Some clever individual must have realized that it would be more efficient to dispense with the tokens, flatten the envelope, and compose the bill of lading out of a set of stamped symbolic shapes strung together in linear fashion. Thus did business interests first give birth to writing in the Western world.

About halfway between Baghdad and Basra, the Sumerian city of Uruk, like the ancient Mayan cities of Yucatán, was dominated by monumental architecture. Separate districts of the city had their own temples, each dedicated to its local protector god. A city god related to one of the aspects of nature ruled over the whole—the god of air, water, the moon, or, in the case of Uruk, Venus, the love goddess. A huge adobe-brick pyramid dedicated to Inanna, and probably to her lost lover Dumuzi as well, once stood in the city center. In the private houses surrounding the largest buildings lived freemen, landowners, nobles, and high officials—among them astronomers and scribes. Most of the surrounding arable land belonged to the temples, a portion of its produce going to the theocracy and its attendant bureaucracy—scribes, accountants, soldiers, fishermen, weavers, brewers, temple gardeners, even snake charmers.

The palace economy was meticulously organized to judge from the tons of pay vouchers and account books that survive in tablet form in addition to the seemingly esoteric astronomical tabulations. But since the rulers were ordained by the gods (I introduced a number of them in our last chapter) and the gods ruled from on high, astronomy was not really so esoteric. For in their movements the stars manifested the very best and worst of human and natural behavior. Whether the king and queen were manifest gods or simply representatives of the pantheon is not completely clear, but archeologists and epigraphers are inclined to agree that palace and temple functioned rather closely together when the earliest forms of astronomy were practiced, although by the time the Venus Tablet was written the ruler may have become more secular.

Still the logic for practicing astronomy in a basically theocratic state was quite simple: If the objects in the sky were associated with the ruler and the pattern of relations could be worked out, then he could know his future, not to mention the future of the market, of potential military undertakings, even whether a marriage ought to be consummated. This interest, coupled with the chronological demands of working out the length of the month and some sort of regular program for fitting months into the seasonal year, mostly for agricultural purposes, provided a twofold need for careful sky watching.

Archeologists excavating Uruk have unearthed the oldest of all the cuneiform tablets. Those at the deepest levels and consequently from the earliest times (they are dated to around 3000 B.C.) mention Inanna, Sumerian Queen of Heaven, and show a star symbol alongside the brief written text (Figure 3-4g). A tablet from a slightly higher level (dated ca. 2350 B.C.) refers to the Underworld Gatekeeper as the star near the rising sun, a likely reference to occasions on which offerings were presented to Inanna in her specific guise as Venus. The special tablets in the British Museum contain mostly astronomical records and omens from the first half of the second millennium B.C. One of them, Number 63 of the so-called series *Enuma Anu Enlil* ("When the Gods of Air and Sky" and named after the first three words of the inscription), is devoted exclusively to the planet Venus. In one spot the text refers to the "Year of the Golden Throne," which Babylonian epigraphers attach to the reign of Ammizaduga, a fairly nondescript seventeenth-century B.C. king of the first Hittite dynasty of Babylon. The Venus Tablet of Ammizaduga has come to be one of the most well studied ancient Middle Eastern documents.

The text, like that on all the tablets in the series, and not unlike the Mayan Venus Table, consists of a monotonous sequence, written in horizontal lines (Figure 4-2): "If X . . . , then Y." In other words, if a natural phenomenon should occur (X in the series can range from a celestial appearance of Venus to the birth of a deformed fetus), then there is a chance that Y will happen in the future. Crop failure, success in business, a military attack, are all appropriate Y's. To give an example from the tablet, "If on the 25th day of the 9th month the Queen of heaven disappears in the east, remaining absent in the sky two months four days, and on the 29th day of month 11, Venus appears in the west, the harvest of the land will be successful."[7]

Even the mathematical statements on Tablet 63, which runs unbroken for twenty-one years, are remarkably like those in the Mayan Dresden document. For example, the real, approximately fifty-day-long, disappearance period of Venus is curiously inflated to an interval that can be as long as ninety days (it is sixty-four days in the example just cited), and the shorter period of absence is equated to seven rather than eight actual days.

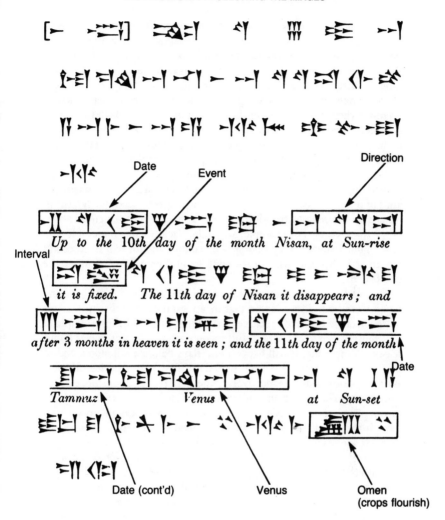

Up to the 10th day of the month Nisan, at Sun-rise

it is fixed.   The 11th day of Nisan it disappears; and

after 3 months in heaven it is seen; and the 11th day of the month

Tammuz   Venus   at   Sun-set

OPPOSITE PAGE:

FIGURE 4-2. VENUS IN A BABYLONIAN CUNEIFORM TABLET

a) A portion of copy K160 from the Library of Assurbanipal of the sixty-third cuneiform tablet of the series Enuma Anu Enlil—the Venus Table of Ammizaduga. b) The names of Venus, dates, intervals, omens, directions, and events are marked out in an enlarged segment, after Sayce (1874). The formulaic repetition is almost exactly the same as in the Mayan Venus calendar in Figure 4-1. (Source: British Museum)

In both cases, the meter of the table is lunar. The scribe only tells us in what day of a particular month Venus disappears and reappears, and he emphasizes the interval between the two stations. The lunar month the Babylonians used, like that of the Maya, could be either twenty-nine or thirty days, and it began on the night when the thin crescent after the new moon was first sighted in the west after sunset. There is no hint either of the year or of the point in the year in which the record is placed. Years as we know them by a continuous sequence of never-ending numbers were not so important to the Babylonians as the regnal years, the time periods that referred to particular rulers, such as "In the eighth year of the reign of King Hammurabi."

Although Babylonian astronomers regarded their sky as basically a kind of lunar clockwork, they did make attempts to keep the months in tune with the seasons by adding an extra month to the calendar about every three years. For example, when the slack became noticeable, the king might issue a decree saying: The coming month will be a second month of Ululu (the name of the sixth month) or Adaru (normally the twelfth).

We have already deduced that those who followed Venus by the moon were bound to discover that its appearances and disappearances on the average backslide relative to the lunar phases. If, for example, a first morning rise occurs on a full moon day, five Venus cycles (eight seasonal years) later it will happen four days before a full moon, and so on.* After seven or eight such periods of five Venus cycles (fifty-six and sixty-four years, respectively), the first rising would once again come close to occurring on a full moon. For such an observer, then, the moon phase and Venus's reappearance would repeat in unison, although the Venus events by this time would have shifted backward through the seasonal year by about sixteen days.†

Modern historians of the ancient Near East have devoted consider-able attention to the problem of dynastic chronology, and Tablet 63, with its precise and detailed astronomical record, has emerged as a seminal text. The chronology question is important not because of anything of note Ammizaduga accomplished but because his predecessor, most famous for having devised the first code of laws, was a member of the same dynasty.

From lists of Assyrian kings and other texts we know that Hammurabi reigned some time between 1900 and 1680 B.C. and that he preceded Ammizaduga by 146 years. A problem for historians who have studied the tablet closely is this: What hypothetical period for the reign of Ammizaduga is compatible with the details of the astronomical record given in the Venus Tablet? The answer to the question What

* This is because five Venus years of 583.92 days are a little less than 2920 days, and 99 full moons are a bit more, as explained in the note on page 95.
† Again, because five Venus years is about two days short of eight sun years.

time period of observation goes with a given ruler? must converge with historical data about relative lengths and sequences of reigns of kings as well as with the absolute data of astronomy. We can back-calculate to within a day the times of full moon three thousand years ago and to within two or three days' observational accuracy the dates of appearance and disappearance of Venus, assuming good weather. King Ammizaduga's tablet is hard evidence in a chronological detective story. While a unique set of real astronomical dates separated by intervals of fifty-six and sixty-four years will match Venus appearances and lunar phases given in the observational record, only one set, the one that best matches history, can give us the truth. Which is it?

Today, after almost a half century of argument and deliberation, the jury is still out, but investigators have zeroed in on 1581 B.C. as the most likely choice for the start of the Venus run of Tablet 63 (alternative choices are 1701, 1645, and 1637 B.C.).

Would the astrologers who cast the omens in clay ever have imagined that we would use their statements primarily to figure out *when* they wrote them rather than *why*? I will be less concerned with the use of the tablet to set historical chronology and more with the observations themselves, in particular with what they tell us about the cosmology of the people who made them and wrote them down and their striking resemblance to those of the New World Maya in the way the sky information is processed and presented.

Unlike the calculations of the Maya, which tracked Venus in time, counting days and placing the planet in a framework of the 260-day ritual cycle, the 365-day seasonal year, and the lunar phase, the primary dimension in Babylonian astronomy appears to have been spatial, in fact, linear—although it is abstract and mathematical and lacks the third dimension conferred upon it by the Greeks. The Babylonian written material that describes the course of the wanderers is a kind of mental arithmetic that tries to keep track of a moving bead on the circular loop of the ecliptic, that imaginary centerline of the zodiacal road. But, like most highway traffic, the bead moves at an irregular rate, slowly increasing then decreasing the distance it covers in a given time span.

Before I plunge into the depths of the tablet, a few words about how the Babylonian astronomers dealt with their sky data. There came a time—we cannot pin it down precisely—when Old World astronomers devised specific models for predicting the course of a planet. They did it first by crude mathematical approximations, for example, by noting that the variable planetary course through the twelve constellations of the zodiac could be conceived in a series of discontinuous steps from one segment of the sky to another. Today's tabloid astrologers still adhere to this stepwise method. For a while, Mars is in Gemini, then suddenly, like a checker on a board, it moves to a new space in Taurus, and so on.

Most Babylonian planetary tables give the position of the object in the zodiac or ecliptic loop in one column, followed by the interval required to get to the next position. This is identical to the format of the Mayan Venus Table, except that degrees (and fractions thereof) on the ecliptic substitute for days. In the earliest Babylonian tables, the added interval appears as one constant over half of the ecliptic and a different constant over the other half, so that the movement of the planet is charted in a stepwise scheme. This fictive description of how a planet moves on its irregular course is a bit like the way we digitally tune a radio or TV. What is in fact a continuous spectrum of transmission frequencies is channeled into compartmentalized units. The celestial channels are the twelve signs of the zodiac.

Later, as greater precision in astrological forecasting demanded, this stepwise scheme was replaced by a uniformly varying interval, increasing over half the cycle, then decreasing over the other half—a zigzag function. Today we set the hour of the day and the day of the year by a fictitious "mean" sun that follows a more sinuous curve. This method is not only more accurate but also more realistic in the way it deals with the sun's turnaround points, where it regresses southward after its northerly course, and vice versa. These appear as sharp-pointed discontinuities in the simpler graphic model. If we failed to keep time by the mean sun, our wintertime hours would differ in length from those of summer; our later afternoon hours would not be of the same length as those around high noon. Better to track an averaged-out sun that lies sometimes ahead of, at other times behind, the one we see in the sky. That nature is not structured in a manner orderly enough to satisfy us is an attitude the ancient Babylonians seem to have shared with us.

Although it may seem crude by modern standards, this method of piecing data together to project the hypothetical place of a celestial body in time and space has the same effect as our scientific way of predicting future events by complex mathematical formulas in the sense that it replaces actual observations with abstract formalisms devised in the mind. Thus, we rely more on our method of judging in advance what will take place and less on the actual sighting of the phenomenon. The observation functions more as a test of the mental scheme, less as a datum that determines an outcome or a result *by itself*. How often do we rely on a scientific weather forecast to decide when to depart, what to wear, or whether to carry an umbrella?

Even though positions and intervals are quoted on cuneiform tablets to the nearest second of arc (1/60 of a minute or 1/3600 of a degree), we should be careful about overestimating the accuracy of the Babylonian observations. Sometimes these notations were the result of applying to the observed data portions of a long-term period or cycle derived by averaging. Let me give a simple example. Suppose you mark off the intervals in days between successive observations of the first visible crescent moon in the evening twilight thus: 29, 30, 29, 30, 30, 29, 30, 29, 30, 30. The average of your ten intervals between observations would be 29.600. Now, you can use this standard interval repeatedly, adding it to the day of any later observation of a first crescent to form a list that will predict when future observations in a series ought to reveal a first crescent. In effect, you will have used a simple formula to produce a predictive table that will be filled with decimals. While you will have carried out your computations to the third place, never will you have recorded a single observation to an accuracy of a period smaller than a whole day.

This lunar interval I chose would do well as a "crescent predictor" for several months. Compared with the period now recognized by modern astronomy, which can be tabulated to six decimal places, it is too long by 0.07 days. This means that if you persisted in using it, actual succeeding crescents would each appear about two hours ahead of our rough predictions. You would be off by a full day in one year. Of course you could use these new data to refine and improve your hypothetical interval, thereby perfecting your prediction technique.

What kind of information is contained in the fifty dates inscribed in the sixty-third tablet in the omen series of the priests of the air and sky god? In this chapter we want to put aside the omens and their application to the royal personage and concentrate on the details of the sky observations. Is the tablet a diary of observations? Was its original version an astronomer's notebook—a celestial historical record gathered by the sky priests? If so, who made the observations? Why observe and record only Venus so meticulously and for such a lengthy period? Why do so during the reign of a single king? Or is the tablet more like the Dresden Venus Table, a model or scheme based on a derived set of formulas that was used to generate predictions about where Venus would be in the future? As one frustrated cuneiform specialist has written, "We have alas no answers to any of these questions."[8]

In fact we haven't even a specific archeological context for the tablet. Nonetheless, it is possible to shed some light on these questions by looking at a sequence of consecutive lines in the table:

> *If on the 12th of the month of Kislev Venus disappeared in the east, remaining absent in the sky two months and four days, and on the 16th of the month of Šabat Venus appeared in the west, the harvest of the land will be successful.*

> *If on the 28th day of the month of Arahsamna Venus disappeared [in the west], remaining absent in the sky three days, and on the 1st of the month of Kislev Venus disappeared [in the east], hunger for grain and straw will be in the land; desolation will be wrought.*

> *Venus disappeared in the west the month of Tammuz 25th; period of absence seven days; rose in the east the month of Ab 2nd.*

> *Venus disappeared in the east the month of Adar 25th, in the eighth year of Ammizaduga.*

> *Venus disappeared in the west the month of Adar 11th; period of absence four days; rose in the east the month of Adar 15th.*

> *Venus disappeared in the east the month of Arahsamna 10th; period of absence two months and six days; rose in the west the month of Tebit 16th.*

> *Venus disappeared in the west the month of Ulul 26th; period of absence eleven days; rose in the east, 2nd month of Ulul 7th, etc.*[9]

The full sequence runs along in this manner, with omens appended only to some of the dates; just as in the Mayan table, these always seem to be linked to the *rising* Venus. There are a couple of brief interruptions and a few scribal errors; for example, two consecutive eastern disappearances are listed, and a morning star appearance after 9 months and 4 days rather than just 4 days of disappearance is an obvious miscue (their clerks made errors too!).

Now, because other documents give the ordering of the month names in the calendar system (Kislev, Arahsamna, Tammuz, and so on), we can compare the statements on Tablet 63 with a real sequence of modern Venus observations. Scholars who have done so generally agree that the material dug out of the earth and dated back nearly four millennia seems to have originated from real observational data,* although there is little doubt that it had been recopied and perhaps tampered with a number of times—always a problem for historians. The scribe who wrote the version we can read today in the British Museum probably knew Ammizaduga no better than you or I know George Washington, even though we both probably have written something about him in our school days.

Inserted midway in the table is the following curious statement: "Once Venus disappears in the west, add seven days to get to the next reappearance and when it vanishes from the eastern sky, add three months to arrive at the date of its return." Other evidence suggests that eight months, five days, a very realistic allotment, was assigned to the appearance interval. We can appreciate the value of this little invention. A clever formula to predict the future whereabouts of

---

* For example, in line 3, a disappearance after morning star interval on the twenty-first of the month of Ab (fifth month) followed by an absence of 2 months, 11 days (2 × 30 days + 11 days = 71 days) does, indeed, lead to a reappearance on the second of the month of Arahsamna (eighth month), as stated, if we assume that each month has 30 days. Just over 8 months later, when we would anticipate a disappearance, the table gives us one. It is recorded on the twenty-fifth of Tammuz, the fourth month (of the next year), exactly 263 days later. These dates and intervals, clearly expressed in the framework of lunar months, look reasonable enough to pass for explicitly empirical data.

the wandering omen-laden light, it stands as a forerunner of the more accurate mathematical formulas that appear in Babylonian science from the time when Nebuchadnezzar (sixth century B.C) became king and restored many of the earlier fallen scientific traditions of ancient Assyria.

Although the focus of the predictions was astrological, these later people were empiricists too, and they seem to have been every bit as attracted to establishing commensurate planetary cycles as the Maya. Their data base consisted of observational records of the planetary positions in the zodiac rather than dates of appearance and disappearance. To paraphrase one of them:

Month 2 Night 7: Mars is above the star Gamma in Gemini by 4 [degrees]

Month 2 Night 23: Mercury is below the star Beta in Gemini by 2½ [degrees]

Month 12 Night 24: Mars is below the star Beta in Capricorn by 2½ [degrees][10]

From these diaries astronomers deduced accurate observable periods for all the planets. Thus they could calculate the multiples of such cycles required for a given planet to return to the same position among the stars.

One text from Persia reads: "Dilbat [Venus] 8 years behind thee, come back— 4 days thou shalt subtract—the phenomena of Zalbatanu [Mars] 47 years, 12 days more. . . ."[11] The first part of the statement looks as if it had been plucked out of a Mayan Venus table; the second part is an obvious reference to the fact that a whole number of synodic revolutions of Mars (25, at 68 days apiece) are contained in 47 years, 12 days.

The essence of this mathematically predictive astronomy consisted of the art of defining how many whole cycles of one set of phenomena accorded with how many of another set. Mayan and late Babylonian astronomers alike were engaged in exploring and discovering commensurate numbers, embedded in nature, that described cycles of recurring phenomena: 99 moons and 5 Venus cycles, 151 Mars cycles and 284 years, the eclipse cycle of 223 moons and 19 seasonal years, and so on.

The most casual observer of Venus, even in the time of Ammizaduga, surely would have been capable of devising a better recipe than the 587-day one. As among the Maya, likely there were other motives behind the choice of this number that remain hidden from us. In Table 4-1, I have listed the fourfold Venus intervals derived from these two Venus texts and placed them alongside the actual intervals for comparison. Notice that the seven-day disappearance, like the eight days the Maya and the Mexicans chose, is pretty close to the real mark, but the

90 days is once again impossibly long compared with the observed 50, or at most 75, days of absence. Recall that the Maya, who tabulated visibility periods of 236 and 250 days, betrayed their motive for selecting them. They did so to fit Venus's motion into the time frame marked by the phases of the moon.

Did the Babylonians habor a similar idea? For example, if we link one of the 245-day appearances in Table 4-1 with the 7-day absence, we get 252 days, which is just 1 day in excess of 8½ months marked by the lunar phases. Also, the too-long 587-day total comes a little closer than the observable Venus cycle to a whole number of lunar months. (It is just 4 days short of 20 moons.) Think of it in the same way as the neat little prescription the Maya devised to keep Venus appearances and disappearances in tune with the moon phases (page 95). You could count out any small aberrations on your fingers.

The modern scientist is as interested in interpreting information as in developing a technique for acquiring and processing knowledge. So, too, the scribe who copied this section of the tablet, or the person who ordered it done, must have been concerned about astrology as well as astronomy. Yet the information in the tablet has as much to do with exploring the mathematical connection between the moon and Venus as it does with foretelling the future of humankind today. Modern

Table 4-1.

Intervals (in Days) Used in Simple Models to Predict the Future Whereabouts of the Planet Venus

Phenomena	Babylonian Model (Ammizaduga Tablet)	Mayan Model (Dresden Codex)	Actual Observable Average
	Days		
Morning Star (visible in east)	245 (8 mos., 5 days)	236	(263)
Disappearance Period (eastern disappearance to western appearance)	90 (3 mos.)	90	(50)
Evening Star (visible in west)	245 (8 mos., 5 days)	250	(263)
Disappearance Period (western disappearance to eastern appearance)	7	8	8 (0–20)
Total Cycle	587 (19 mos., 17 days)	584	584 (583.92)

investigators have been able to correlate the types of omens centered on the rising Venus with the names of particular months in the calendar and in turn with everyday functions and needs of farmers in ancient Iraq. For example, the tablet says that good harvests accompany risings in the months of Tebit, Tesrit, and Sabat; that positive rapport between kings comes in Sabat and Adar; destruction will happen in Ab, Arahsamna, and Kislev; floods will occur in Ayar, Tammuz, Ab, Arahsamna, and Sabat.

The temporal architecture of these predictions is neither random nor whimsical, for if we map out the placement of these omens across repeatable twelve-month cycles, we find that statements concerning crops, such as information about the harvest and heavy rains, fall in time periods when farmers would be in dire need of such knowledge. Omens on water flow occur in the crucial growing period May to July (with one reference to March), whereas omens on the quantity of the harvest begin in July and end in the early winter months.

Some documents of mundane commercial concern have survived that may have a bearing on interpreting what we read in these Venus predictions. Apparently croppers would sign contracts promising the delivery of a harvest of barley, one of Mesopotamia's staple crops, over a particular range of seasonal dates. The modern harvest date in Baghdad falls between mid-April and mid-May. Keeping in mind that omens are statements of anticipation, it would be logical that the Venus rising that would offer information of greatest need to the farmer—such as yield, quality, or tardiness of the harvest—might be expected to fall in the months before harvesttime, and this is exactly what we find in the tablet. However, because other crops, such as dates, were harvested at a different time of year (September to October), we must be careful not to confine all predictions about fertility to one time of the seasonal year.

By contrast, political predictions are mirrored in Venus risings that fall on the average in a different part of the year. For example, all good relations between kings are recorded in our monthly equivalents of November to January. Evidently, when all attention must be devoted to cultivating the land, one has little time for bickering over who shall administer it. Like the Maya, the Babylonians had their war seasons.

One historian of science has characterized the Venus Tablet of Ammizaduga as "primitive and, to our way of thinking, of little significance."[12] But I think it might be more constructive to think of this slab of clay as a statement emanating from an age when basic human needs were the same, though the ways of dealing with nature were a bit different—an age when planetary motion was reckoned in a more direct, less abstract way.

My goal in dissecting the Venus Tablet of Ammizaduga and the Venus Table in the Dresden Codex has been a fairly modest one. I wanted to bring to the surface a sense of the rigor and precision that lies at the foundation of what many of us might casually dismiss as prescientific astronomy—a kind of simpleminded activity in which our muddleheaded predecessors had engaged before they became wise enough to separate meaning in their lives from what they observed directly in nature, and skilled enough to construct mechanical, self-regulating theories to explain not only what already had happened in the sky but what was yet to come. This theory of human history is just too shortsighted.

I find it remarkable that two cultures on opposite sides of the globe chose to structure Venus events in an almost identical way. Nor am I about to embark on the ship of transoceanic diffusion, for the cycles of time manifest in the heavens, though they be many in number, are plainly visible the world over. That moon cycles would be used by different viewers as the fabric on which to embroider patterns of the course of the sky's brightest star comes as no surprise to anyone who watches nature.

But why did they go through all the trouble? Why the detail? Fear alone could never have motivated such a course. I suspect the Babylonian priest who invented the 587-day Venus formula possessed an innate desire, what we might call a natural curiosity, to establish an orderly and unified method of making future predictions based on past observations. The same held true of the Mayan astronomers. They could not have escaped wondering why they sometimes needed to deduct four and at other times eight days from the Venus count to keep their canon of predictions and ritual day names on target with the rising Venus. To confine conjecture, meditation, and speculation only to the present and to ourselves may reflect our own insecurities.

But let us neither misrepresent nor romanticize the roots of this innate celestial curiosity. Ancient desires to explore the heavens clearly were couched in a religious framework. No one believed in a mechanistic universe governed by gravitational principles and operating with no special concern for people, animals, and plants. Such a worldview would have made about as much sense to the Babylonian or Mayan as painting in the Cubist style, rock music, or a democratic society that emphasizes the role of the individual.

The folded bark book that New World sages carried from town to town to confront their clientele and the Old World clay tablet that perhaps once stood in an ancient Middle Eastern library, each having been copied and recopied, provided centuries of grist for astrological mills. The twenty-one consecutive years of Venus observations recorded on Ammizaduga's tablet constituted a set of basic data, and whatever observations lay behind the Dresden Venus Table (unfortunately not a single Mayan astronomer's notebook has survived the Spanish purge) also served as valuable information for seeing into the future, just as our census statistics stored in computers generate predictions on possible future courses of national action on the world food and hunger problem.

The process of following the planetary wanderers I have been discussing in this chapter was scientific astronomy without a doubt—precise and predictive, even if it rested on an observational data base quite different from what we are used to. If we have learned anything from this and the preceding chapter, we should be aware that yet another confrontation awaits us. For if we would seek to understand why the ancient astronomers insisted upon carrying on such rigorous programs as these two Venus texts suggest, then we are forced to open the back door of the closet of science and engage their astrology. The Maya made the Venus Table because Venus in the sky was the manifestation of the influence of Kukulcan, a power reflected in the divine ruler-king who was linked to the sky god by blood lineage. And the Babylonians constructed the Tablet of Ammizaduga because Venus-Ishtar was an arm of the celestial triad whose changing meaning was expressed directly to them in each of her comings and goings. Of all that we have learned from the complex planetary tables that have been unearthed by archeologists and saved from destruction by manuscript collectors, we have the greatest difficulty coping with the bond we severed between nature and humanity that seems somehow to have remained intact in the minds of our predecessors. We almost envy them for it. Can we really imagine what the public comprehension of the esoteric writings of astronomers would have been without the relevant astrology to bind the products of ancient sky science to the everyday world?

What was true in ancient Babylon or Copán also would have held in classical Athens. The writings of Aristotle or the dialogues of Plato that tell of the way the spheres above turned could, taken by themselves, scarcely have been understood by fisherman or shepherd. This was powerful intellectual stuff with virtually no practical application, nothing to satisfy either common citizen or slave.

Are the ancients so remote from us then? With a little probing, I have found in their endeavors a surprising number of scientific traits like our own. Observation, quantification, even mathematical formulation lie on one end of the chain link between the mystical and scientific ways of thinking. But what lies at the other end, so oft immersed in occult shadow? Is there a more common ground on which to place astrology so that we modern nonbelievers can share in its meaning? Can its methods be articulated logically? Must we always revert to keeping those parts of the ancient worldview that we value and discarding the rest? What motivated the ties that bound their way of thinking about the natural world and why do vestiges of their way persist even to this day? These are all important questions that will be explored in the next chapter.

CHAPTER 5

ASTROLOGY:

BELIEVING IN THE IMAGES

> When the sunset is inseparable from the thought of
> death, then dawn is a sure sign of resurrection.
> —HENRI FRANKFORT, *ANCIENT EGYPTIAN RELIGION*,
> 1948, P. 29

WHERE OMENS COME FROM

Today the sunset belongs to nature, and death belongs to all of us. There can be no association between the two. But for most of the history of the world, people have responded to nature differently. Each of her signals carried a message— one to which they needed to reply rationally and with appropriate action. That people can act of their own will— that they can be active participants rather than the mere objects of superiors' commands is a quite novel perspective.

Up until the past three centuries of West European history, the belief in astrology, literally the "doctrine of the stars," had been practically universal. It penetrated all levels of stratified society, from nobility to peasantry, and directed all forms of activity, from politics and science to medicine and agriculture. It was part of life. Belief in astrology in Western culture began to decline in the seventeenth century. It was not swept away by scientific heroes on horseback, who suddenly saw the light of empirical, objective truth. The death of astrology has been a slow and agonizing process, today very nearly complete. In a handful of generations, tracing the effects and influences of the planets on mundane affairs has been reduced from exalted art to pseudoscience. Although we have dissected astrology from astronomy, if we really care to follow the changing dialogue between people and planets through the ages, we are compelled to know what astrology was like in its own time and how it has been reduced to what it is today.

Astrology is a form of divining—looking into the future by consulting the superior forces embedded in various entities that make up the natural world. Reading the innards of animals was one common form practiced worldwide—ancient Roman diviners examined the sheep's liver, the Incas the inflated lungs of a llama or the stomach of a guinea pig. The Greeks watched patterns of birds in flight, the Maya cast kernels of corn, and medieval diviners could see your future by gazing into a cup of your freshly expelled urine. Looking for signs was a way of clarifying your relationship to the powers of nature—to provide a guide for moral action, to create order in an uncertain future.

How does astrology work? To begin with, there were different forms of it, each developing at different times and places and to varying degrees. Natural astrology, for example, was the study of how the arrangement of the sky and all that is in it could be used to influence the occurrence of winds and storms. (As I shall show later, *influence* is the key word here.) It seems perfectly logical to us that the sky condition can predict the weather. What sailor, however landlocked, hasn't heard the rhyme "Red sky at morning, sailors take warning/Red sky at night, sailors delight"? And who hasn't seen the halo around the sun or moon? Halos were called garlands by the Romans, and Virgil regarded them as signs of the tempest, because the wind generally begins to blow from that quarter where the circle of the halo starts to break.

The sixteenth-century puritan philosopher William Fulke of London tells us in his *Book of Meteors* that rings around the sun and moon— like rainbows, mirages of castles in the air, night suns, and double moons—do not occur by chance. Instead, they are sent by God as "wonderful signes, to declare his power, and move to the amendment of life."[1]

By contrast, judicial astrology pertained to foretelling the lives of people and nations. Genethlialogy was the specific practice of casting a nativity horoscope—the one we still find in the daily newspaper— based on the positions of the planets at the time and place of your birth. This most familiar form of astrology comes down to us from the Chaldean world of Mesopotamia in the second millennium B.C. Regardless of how we might categorize astrology, the question about the dimension of human freedom will arise. Do the stars rule us or do they merely indicate a predilection for future possibilities—the way your genetic code might make you prone to certain ailments, which you can astutely avoid or delay by medication or proper diet? Different cultures offered different answers.

At the root of the old Chaldean system lies the belief that the destinies that control us are the result of the complex interaction of celestial spirits, whose influence touches us through the energy their rays give off. We mortals here below—in the *sublunary* ("beneath the moon") world—vibrate sympathetically in response to these celestial emanations. The radiation can increase or be weakened, according to where the celestial luminary or source lies in the sky and according to the time of year or the time of night the emanator rises or sets. The space of the zodiac, along which these influential emitters travel, is divided into twelve parts, a choice that derives from the number of full moons that could be fit most closely into the seasonal year. In this system every category of existence, like every division of time and space, can be placed under some influence or rule and, therefore, be subject to extreme categories of behavior—beneficent or tyrannical, strong or weak, positive or negative—of which transcendent gods, like ordinary mortals, are capable.

In early Assyrian Babylonia (1700–1400 B.C.), as among the Classical Maya (A.D. 200–900), the places of the planets in the zodiac were employed in a relatively simple cyclic scheme to give short-range predictions, generally of worldwide impact. Predictions as well as explanations usually centered on sudden and striking events that would affect just about everyone—a tidal wave, flood, or earthquake; a major turn of events in warfare; or, as in these two instances (the second one after the fact!), a sudden plague on the cattle of an ancient city or on the entire European population (the Black Plague):

> When a halo surrounds the Moon and Mars stands within it, there will be a destruction of cattle; [the city of] Aharru will be diminished. (It is evil for Aharru.)[2]

> On 20 Mar. 1345, at 1 P.M., there occurred a conjunction of Saturn, Jupiter and Mars in the house of Aquarius. The conjunction of Saturn and Jupiter notoriously caused death and disaster while the conjunction of Mars and Jupiter spread pestilence in the air.[3]

Where does power come from? The power of one individual over another, of a ruler over his subjects, of people over nature, or of nature over people? What is the source of authority, control, and dominance? The Classical Maya had an answer, not too different from that of the Babylonians. It began with the faith that we all live in an animate universe. Heart and mind, eye and hand, all parts of life vibrate together and influence one another—harmoniously when the organism is in working order and in disharmony when things go awry. When the sacred spirits of the Upper- and Underworlds move, we here in the middle on earth are profoundly affected by their actions. If we pay attention, the influences can become sensible to us in the shifting of the wind, the falling of the rain, or the movement of the stars. And the way we behave, in turn, can affect the way they act. The Maya shed their blood in sacrifice to open a channel of communication to those animate powers—to pay their debt to them.

This celestial power brokerage, forged by the blood currency between the people and the gods of nature, was all pretty clearly laid out in Maya foundation mythology, as I described in Chapter 3. After the dawn of creation, the *Popol Vuh* story says that the first humans, ancient ancestors of the Quiché Maya, went before the representatives of the god of fire, Tohil, to negotiate for his staple element. They asked, "What can we give Tohil in return? What about metal?" "We don't want metal," said his spokesman. "When the time comes they'll be suckled on their sides, under their arms, Tohil says. Let all tribes be cut open before him . . . and their hearts be removed through their sides, under their arms."[4] Exorbitant terms? You need to give in order to receive. The people agree to heart sacrifice and they get their fire.

In rule by divine kingship, kings and queens are bound to the people by the belief that they themselves are sacred beings, individuals kept safe above all ordinary mortals because they alone are the manifestation in human form of the spirits that rule an animate universe. We can understand why Chan Bahlum of Palenque, for instance, donned one of his godly impersonation outfits and let blood before the assembled audience in front of his temple on the night of the great conjunction described in Chapter 4. Through the bloodletting rite, the king was reaffirming in the present what had been ordained in the *Popol Vuh*. His act reforged the vital link made by his ancestors with their cosmic kin, the Palenque Triad of deities, Tohil, and the other deities who populated the living universe in the form of the wandering sun, moon, planets, and stars.

I have the sense that heaven was not so far away in the eyes of these people, who believed that individual, civic, and social truths could be revealed by consulting with nature: the shifting of the wind, the arrival of the rain, the appearance of the stars.

A Babylonian priest pleads poetically through the animal he is about to offer to allow him to see into the future:

> *O Great ones, gods of the night,*
> *O bright one Gibil, O warrior Irra,*
> *O bow [star] and yoke [star],*
> *O Pleiades, Orion and the dragon,*
> *O Ursa Major, goat star, and the*
> * bison.*
> *Stand by, and then,*
> *In the divination which I am making,*
> *In the lamb which I am offering*
> *Put truth for me.*[5]

When a living universe is your home and all parts of your world pulse harmoniously, then you talk to the stars, and they talk back to you. When planets come together, your fates come together with you. Truth is revealed to you not from the instruments of science, not from computer, telescope, or mathematical formula, but from amulet, incantation, and exorcism.

"Put truth for me," in what I offer, the poet beseeches his celestial gods. Like the Maya, the Babylonians believed in two planes of reality with a human mediator in between. We were put here on earth not just for ourselves but for the express purpose of interpreting the actions of the omen-bearing invisible spirits who speak through the entrails of sacrificed animals and aborted animal fetuses, who empower the birds to fly and the planets to roam across the sky:

> *When a hermaphrodite is born which has no . . . the son of the palace will rule the land [or] the king will capture.*

> *When a fetus has eight legs and two tails, the prince of the kingdom will seize power. . . . A certain butcher whose name is Uddanu has said, "When my sow littered, [a fetus] had eight legs and two tails, so I preserved it in brine, and put it in the house."*

> *When it thunders in Tisri, there will be hostility in the land. When it rains in Tisri, death to sick people and oxen [or] slaughter of the enemy.*[6]

These omens were written in Babylonian cuneiform notation upon clay tablets over 3500 years ago. All follow the basic formula: When X happens in nature, Y will happen to us. Today we, who think of mind and matter as separate and unrelated, cannot be expected to see absolute truth in such formulations. Our disconnection with the natural world guarantees it. We might become awestruck by witnessing an eclipse or an exceptional sunset, or terrified by nature's power when caught in an earthquake or thunderstorm or knocked over by a pounding surf, but we are scarcely capable of being moved with such force that we would speak directly to an imagined power immanent in nature: "Put truth for me!"

Whether on clay tablet or parchment codex, these omens reveal that people once saw themselves in an environment that was alive and a part of them—a place with both persona and sentience. The ancient Greeks showed this belief: "The moon . . . bestows her effluence upon the earth; for most things, animate and inanimate, are sympathetic to her and change together with her; the rivers increase and decrease their streams with her light, the seas turn their own tides with her rising and setting, and the plants and animals . . . become full or diminish together with her." This passage from Ptolemy's *Tetrabiblos*, written in second century A.D. Alexandria, sounds logical even to the astrological nonbeliever. The ebb and flow of the tides follows the moon, the opening and closing of flower petals follows the sun of day. The rising and falling of the river, the fruitfulness and barrenness of the land, and of cattle and wild animals—all of these processes follow the sun. If the action of the moon can be correlated with the tides, then what about the tides in the affairs of humanity? How far can the logic be stretched? Why not imagine that other influences and virtues are hidden away in the universe and that these powers can act both sympathetically and antipathetically—ebbing and flowing like the tides—toward events that take place in the realm of terrestrial affairs? It all seems perfectly rational.

In a sensate universe, it is logical to suppose that the most influential predictors of our destiny would be the very ones who behave most like us: petulant, unpredictable, always changing their minds—the planets. They were the most diverse, fickle, hard to pin down for astronomers. They sway to and fro, stop dead in their courses, turn backward, then spring ahead again in double time, waxing to brilliance only to fade away as they march across the zodiac. Often they cross paths, even confront one another. They seem almost human. Remember that when the Chaldeans uttered their names: Ninib (Saturn), Merodach (Jupiter), Nergal (Mars), Ishtar (Venus), and Nebo (Mercury), they seem not to have been addressing abstract personages who guided the forces of nature in a metaphysical way. What happened there in the sky was not the effect of an inalterable set of causal laws but rather voluntary acts committed with the same sort of intent and backed by the same sort of intellect with which any mortal down here on earth might be endowed, except much better honed and developed.

Old wine in new bottles—we are constantly acquiring and reshaping the myths and legends of our forebears. The ancients were no different. The old stories and the gods who starred in them were never completely discarded over the aeons but merely garbed in new dress.

The lending process gives us our sense of tradition, whatever we perceive as our ways, our customs, which we can trace back far enough to legitimize our goals for the future by placing them upon a firm foundation in history. Practicing Christians reenvision stories in the Bible to exemplify the just moral principles of the twenty-first century—not necessarily those of the time of Moses or Christ, which of course we can never truly experience. Interpreters will disagree on the extent to which biblical characters and story lines have been altered and recast, but few will deny that some restructuring has taken place. For example, human dominion over the animal world expressed in Genesis is taken by many today to imply a custodial relationship rather than domination, which has come to have a negative connotation.

Any culture that claims descent from a predecessor has at least one foot firmly implanted in the mythology of the donor culture—whether it be the Aztecs who borrowed from the Olmecs, the Greeks who borrowed from the Babylonians, or the early Christians who borrowed from the pagan astrologers.

In Chapter 3 I explored some of the logic behind the assignment of various qualities to specific planets. Nebo became a prophet because he always stayed so close to the world and repeatedly dipped down into the human realm; therefore, he knew what was happening everywhere, and he never hesitated to give his opinions on every subject. As Greek Hermes he came to govern all learning preserved in the form of the written word. Later, as Roman Mercury, he revealed his mercurial temperament, which for us still means lighthearted, flighty, changeable, hard to pin down.

I showed how Venus-Ishtar's extreme swings of position between evening and morning star made her capable of the lowest, most reprehensible debauchery as well as the highest form of pure love. She carried the full spectrum of femininity as conceived in the male Chaldean mind; she became the celestially personified role model, good or bad, in the extreme. And we still can feel the dull, heavy, gray tones of the saturnine one if we carefully watch him plod slowly along the highway of the stars, or the fiery red, warlike character of the martial deity. These sky gods emerge as characters not unlike our modern movie, TV, or comic book heroes whose human qualities—sexuality, goodness and evil, strength and prowess—are magnified to extremes. Like the superintellects and super egos our cartoonists create for the screen, they populate an imaginary plane of reality, a medium in which it becomes safe to raise questions about our own existence that we do not ordinarily discuss: Where does the power of human life originate? Does life survive and continue after death? What is real love?

Just as the planetary gods mirrored people's images, so was the pantheon a reflection of the whole of human society. For example, the divine aristocracy that constituted the Chaldean enclave of deities took on the same highly structured hierarchical plan as the society that worshiped it. Sometimes these celestial gods were the spirit or the ruler of what they represented rather than the material itself. Anu was the essence of things in the sky, Enlil that of air, Enki of water. They variously lived in their domain or within its house. Like the people who created them, the gods became socialized. Occasionally, the god of earth traveled the region of the zodiac to inspect its stations. The spirit of Venus mixed with that of Mars. Like Mayan Tohil, god of fire, the Old World gods appointed deputies and ordered the construction of divine oracles, amulets, and fetishes to facilitate the dialogue with people. Those on earth who advanced that dialogue were accorded particular social standing.

Being an astrologer in ancient Uruk, Palenque, or Luxor was probably a bit like being a respected heart transplant surgeon in the present turn of the century. An Egyptian inscription on a statue of the astronomer-astrologer Harkhabi describes him as

> *hereditary prince and count, sole companion, wise in sacred writings, who observes everything in heaven and earth, clear-eyed in observing the stars, among which there is no erring; who announces rising and setting at their times, with the gods who foretell the future, for which he purified himself in the days when Akh [an unidentified celestial body] rose heliacally beside Benu [Venus] from earth and he contented the lands with his utterances.*[7]

But if his client were king or pharaoh and serious affairs of state were involved, an astrologer's job could become as demanding as that of the secretary of state. A surviving message from seventh-century B.C. Assyria reveals the scribe Munnabitu's stress as he laments the difficulties of his job:

> *The king has given me the order: Watch and tell me whatever occurs! So I am reporting to the king whatever seems to be propitious and well-portending [and] beneficial for the king, my lord [to know]. . . . Should the king ask, "Is there anything about that sign?" [I answer], "Since it [the planet Mars] has set, there is nothing. . . ." Should the lord of kings say, "Why [did] the first day of the month [pass without] your writing me either favorable or unfavorable [omens]?" [I answer], "Scholarship cannot be discussed [heard] in the market place!" Would that the lord of kings might summon me into his presence on a day of his choosing so that I could tell my definite opinion to the king my lord!*[8]

Yet another astrologer remarks openly about his own failure to note a last disappearance of Venus in the month of Kislev:

> *When Venus in Kislev from the first to the thirtieth day disappears at sunrise, there will be famine of corn and straw in the land. The lord of kings has spoken thus, "Why has thou not [observed?] the month, and sent the lucky and unlucky?" The prince of the kingdom has been neglected, has not been obeyed. May the lord of kings when his face is favorable lift up my head that I may make decisions and tell the king, my lord.*[9]

Why the errant scribe had failed to apprehend the vital information the white luminary was ready to convey to her worshipers we can only guess. Did he pull out the wrong list of risings of Venus on different days of the month and then disregard the corresponding events on earth written against them? Or was he simply too overworked and exhausted to perform his task?

These poignant quotations, still alive on clay, portray the astrologer not as we might be inclined to see him, a charlatan, but instead as a helpless spectator very hard at work attempting to follow the rules of his profession. He scarcely seems an arrogant possessor of true knowledge—a wielder of real power—at least not in the eyes of his king. In some cases, the word from on high seems clear enough, but in others—"Put truth for me"—it is rather opaque. But then what successful religion does not possess the unanticipated, the novel? Unlike science, religion is not in the business of giving definitive, precise answers to all the questions we can pose.

Figure 5-1 offers two real-life portrayals of ancient astrologers from opposite sides of the world. Whoever they were, these careful observers and omen writers, often working in teams, evidently were well kept and provided for by the ruler. Those from the Middle East appear to have been no different from the astrologers who served the kings of Palenque and Copán.

And New World vase paintings such as the one depicted in Figure 5-1a portray specialized schools of omen bearers. Highest ranked were the astronomers who made the observations. Next came the mathematicians who did the calculations, then the codex painters who produced the documents—note the ink pot next to the artist at work with brush in hand. The lowliest class was probably the assistant who measured out the spaces on the bark paper for the glyphic block and numerical dots and bars that made up a codex. A recently excavated tomb at the ruins of Copán revealed remains of the son of a king thought to have been one of these high scribes. Paint pots and offerings to the Mayan god of scribes were dug up next to his bones. No statements such as nervous Munnabitu's survive on this side of the Atlantic—only the words of the Spanish Roman Catholic chroniclers, like those I quoted on page 93, who, even though they saw fit to condemn the documents they wrote, exalted the wise men who created them.

Given the detailed terminology they employed to describe what happened when celestial bodies lay near one another, the Chinese astrologers seem to have been far more preoccupied than their Western counterparts with planetary conjunctions. Every planetary contortion on the starry background was explicable as a reverberation of some moral aberration in the governance of the state. There was no need for physically based laws, no motive to search for independent planetary orbits such as Galileo and Copernicus would later attempt in the West. According to Chinese astrology, planetary essences descended upon humankind during planets' advances and regressions as well as their rapid movement through the constellations of the Chinese zodiac. For example, the action terms *chou*, *ch'eng*, and *ling* referred respectively to one planet passing around another, descending from above, and moving upward from below. Other terms signified when one planet rushed by another or concealed another, or when two moved in opposite directions along the same line, when they covered each other, joined then separated, or hit each other.

Figure 5-1. WHO WERE THE ASTROLOGERS?

5-1a. Healthy-looking Mayan scribes painted on a vessel are shown writing a codex. The one on the left and the third from the right may be teachers. The latter is shown, pen in hand, hovering over a text conversing with his students. Notice the Mayan dot-and-bar numerals coming out of his mouth. The inscription across the top tells whose vessel it was. (Justin Kerr, *The Mayan Vase Book,* New York: Kerr Assoc., 1989, 67. File No. 1196.)

5-1b. This staff of a government officer in the time of the Memphite dynasties of Egypt includes a pair of scribes seated at their desks. (Drawing by Faucher-Gudin in G. Maspero, p. 289.)

Chinese astrology boasted an internal logic to these active encounters, one that reflects the Chinese penchant for mixing complementa-
rities. To explain a few cases, conjunctions of Venus and Jupiter were
called *p'in mou* (female and male). When Venus (the metal planet) and
Mars (the fire planet) combined, there was fusion (*shuo*); when Mercury
(the water planet) encountered Mars, there was tempering (*ts'ui*). Saturn (the earth planet) coalesced with Mercury to give a blocked channel
or *yung ch'u*.

The sun was the essence of *T'ai Yang*, the force of the sustenance of
life, benevolence, and virtue, all obvious qualities of the emperor. But
it was also capable of revealing his shortcomings. When sunlight was
dim from noon to sunset, his legislation was too oppressive. When such
dim light coincided with the sounds of crows, he misgoverned. Any
changes in the sun's appearance presaged a change in the state of the
empire. In times of war, for example, a change in the sun's color could
augur defeat in battle, but in times of peace it might signify the death
of a noble. The moon was the sun's counterpart—not its opposite but
its complement. Its essence was *T'ai Yin*, a force associated with the
empress and her qualities, but it too bore watching, for under a wise
empress it kept its course. It moved south or north when the rules of
punishment were not correctly applied, and a sudden change of color
meant the empress had behaved foolishly.

The universe revealed to us by the ancient Chinese astrologers seems
to have had a sound moral structure. True, it could be perturbed by
signs from above, but basically, when the world was in equilibrium, it
was not such a bad place. Political scientist Andrew J. Nathan writes:

> *This view . . . has roots in early Chinese thought in which the human
> world has been seen as an integral part of a cosmos ordered by moral rather
> than physical consistency. Cosmic harmony was thought to descend from the
> heavens to the natural world to human society and finally to the mind of
> the individual—and then to extend back, for the Chinese saw the entire
> cosmos as responding to the moral or immoral behavior of human beings,
> especially the ruler.*[10]

The "view" of which Nathan speaks concerns the potential of modern
Chinese culture under the democratic regime espoused by the Beijing
students' manifesto issued in the failed revolt of the summer of 1989.
Whether the faith in a morally sound, harmonious universe would still
yield a smooth transition from totalitarianism to democracy may remain
untested. But what I find interesting is that ancient China's mental outlook on the universe has persevered for hundreds, even thousands, of
years and can surface in a discussion of contemporary politics. The unity
of cosmos and humankind was indelible in the history of China, and
what has been written about Chinese beliefs in the planets proves it.

GIFTS FROM THE GREEKS

T he astrologers of Old World Babylonia and Egypt, China, and New World Yucatán seem to have been much alike— highly specialized in the mathematical skills associated with daykeeping yet schooled in ways to tease the future out of the stars. The form of astrology our modern culture has acquired descends, oddly enough, from the same civilization that gave us its appreciation of the beauty manifested in the sculpted form of the human body, of the stateliness we have so long sought to imitate in our architecture, and of the sturdiness of its form of government, so well articulated by Aristotle in the *Politics*. If we really want to know how astrology became disentangled and finally estranged from its sibling science of astronomy, we need to cast our eyes to the ancient Aegean.

If *zodiac* is one of the words that comes into your head when I say *astrology*, then a second word must surely be *horoscope*. We owe our word *horoscopus* to the Greeks. It means "I observe the hour," or colloquially, "I watch what rises," and it refers to the art of predicting the general patterns that are preprogrammed in your future life based on an examination of the celestial bodies that were coming over the horizon in the east at the time and in the place you were born. Unlike the Babylonians, whose astrology was concerned with what might befall an entire state depending on what was happening in the sky, the Greeks— reared in a democratic system—believed that everyone should have his or her own personal horoscope. Everybody had a right to knowledge about the future.

The French anthropologist Jean-Pierre Vernant has brilliantly outlined the intellectual revolution that advanced the idea that we live in a space-bound universe of circular orbs. He traces it to the realm of politics, specifically to the birth of the Greek *polis*, or city-state. With the sack of the Persians at Marathon in 490 B.C., the Greeks opened an era of freedom. They had rid themselves once and for all of the yoke of Eastern barbarism. Now they could turn their thoughts to expansions of their own, to advancing trade and commerce, and to probing and contemplating the workings of a much bigger world than the tumultuous one that immediately surrounded them—the one from which they had just extricated themselves.

One distinct advantage of highly structured, specialized societies such as Greece in the fourth century B.C. was the leisure that their economy afforded them—leisure that was not spread out over the middle classes as we perceive it today but concentrated only in the highest social strata—the wealthy landowners and nobles. When you have a lot of time on your hands, you can afford to engage in esoteric pursuits.

You have the luxury of wondering where dreams come from, how the underworld is constructed, what makes the stars move about the sky.

What we interpret to be the highest forms of astronomy—the most mathematically articulate, the most precisely predictive—flourished in these sorts of societies. The ancient Greek and Islamic, the Mayan and the Babylonian civilizations molded and shaped intricate astronomies and the astrologies that went along with them. And by not being limited by everyday absolute needs—not needing to tend his cattle, wash his clothes, feed his family—the elite astronomer-astrologer was free to pursue his craft to the limits of his own intellectual curiosity. Some were naturally more curious than others. This probably explains why the Dresden Codex is far more accurate in its astronomical predictions than it needs to be. It also accounts for the ostensibly cabalistic nature of the astronomer-astrologer's endeavors. Who but the initiated few, those with extensive training and experience, could ever hope to understand the complex subtleties of making the computations, and casting the particular omens, that took place in the privacy of the temple? Our science may not be so different, at least as the public perceives it. What happens in the lab is rarely comprehended by people outside the inner sanctum of those who understand the function of a single piece of equipment in the experimental train. Expert journalists struggle to bring the fruits of the scientists' labor to a public hopelessly unprepared (quantitatively innumerate and scientifically illiterate, to use the jargon of the present day) to understand the meaning of the work of their high priests. Classical Greece could not have been so different.

Our gifts from the Greeks also include the idea that good and bad times alternate. Lucky and unlucky days and months were marked out by the motion of the sun and moon along the constellations of the zodiac. Attaching a person's birth date to the predictive scheme led to the development of the sort of character analysis that comes down to us in the form of newspaper astrology. But Greek astrology was not pop astrology. It was part of rational science and philosophy, wedded to the Greeks' idea of the cosmos as an ordered whole. The roots of their spatial model of the universe were closely tied to their ideas about how society functioned. Greek town plans were rectangular because a right angle was the only rational angle, and city space was divided according to the class of inhabitants (warrior, merchant, and so on) and whether the land was public, private, or sacred. Greeks never separated real life from physical space.

If they could secularize and organize the city, why not the universe? In the city in the sky, the Greek philosophers looked for and began to see things that seemed to function in the same way as in their ideal city-state here on earth. Just as urban structures clustered about the open public space of the *agora*, or marketplace, the planets seemed to pivot at various distances from a center, the earthbound Aegean world.

When they spoke of the earth as the center of all things, the Greeks called it the "hearth of the universe," the center of orthogonality of the family, the point of contact between earth and sky. Our very word *focus* is the Latin term for the hearth, which once was the center of suste- nance, dialogue, and religious practice. It housed fireplace, kitchen, and altar in the same place. The geometrically organized universe, a concept held from the time of Pythagoras all the way to Ptolemy, argues Vernant, is a projection onto the world of nature of the great social model of the human community developed in Hellenistic times.

Geometry—the word means "land measure"—is perhaps the greatest intellectual gift of the Greeks. It has gripped our imaginations and shaped our philosophy about the structure of the universe ever since they offered it to us 2500 years ago. In the Greek colonies of southern Italy, the Pythagorean school of philosophers developed a set of underlying principles by which they believed the universe functioned. Their ideas were rooted in numerical ratios and geometrical proportions. Their highly abstract, theoretical concept of the universe is a long way from the animated, organismic view expounded by the earlier astral religions of the Middle East or by the Ionian school of philosophers. Its real unique- ness lies in its space-centeredness, for the Greeks were preoccupied with the organization and movement of things through space.

Why, for example, should the earth remain fixed at the center of the universe? Because it lies at the same distance on all sides measured from the star sphere. Therefore it has no reason to move to the north, south, east, or west. So nothing from the outside can hold power over the earth, for all power resides at the center.

The history of Western astrology, then, is part of the history of the social revolutions that moved the science of star predictions toward and then away from a dependence upon precise observation. There were times, such as the Alexandrian period of the first few centuries B.C. and A.D., when astrological prediction became especially closely tied to perceptible natural phenomena that almost by themselves seemed ca- pable of transmitting meaning. But at other times, such as during the early Roman-Christian period, as I shall show, societies developed as- tral mythologies that appeared quite remote from direct observation of the phenomenal world.

A visitor to modern Alexandria, Egypt, where the Nile flows into the Mediterranean, would be hard-pressed to believe that, from the time of its namesake, Alexander the Great (fourth century B.C.), until the Rome of Antony took over the Egypt of Cleopatra in the first century B.C., this was the intellectual capital of the Mediterranean world. Today it is a run-down seaport with polluted skyline and a seafront cloned with high-rise apartments. It is a city with no real national character—an odd mixture of Middle Eastern and European peoples. The harbor is no longer lit by the beacon of the great lighthouse, one of the celebrated Seven Wonders of the Ancient World. Other Alexandrian lights were snuffed out long ago. Its museum, literally a place where people gathered to study the muses, was one of the finest facilities for scholars of the time, a sort of government-supported research institution. Here also once stood the world's greatest library. A magnificent structure housing more than half a million manuscripts, it was burned in a religious riot by early Christian fanatics.

Alexandria originated the greatest advances of mathematics and along with them the creation of an earth-centered model of the orbits of the planets sound enough to survive fifteen centuries. And here, too, the science of astrology flourished, for the very creator of a new solar system model, Claudius Ptolemy, who wrote ancient astronomy's most encyclopedic work, the *Almagest* (a later Arabic corruption of the Greek word meaning "The Greatest") also composed astrology's bible, the *Tetrabiblos*. But then, the Greeks would not have differentiated between the two, classifying both works under the heading *astrologia*.

At a personal level, we know nothing about Ptolemy except that he was alive between about A.D. 85 and 115 and that his name ties him to the dynasty that ruled in Egypt at the turn of the first century A.D. His *Tetrabiblos* (literally "Four Books") advocates the careful determination of the appearance of the course of heavenly bodies and the codification of mathematical laws to "save the appearances"—to describe, delineate, and predict where they will be at any future time. But at the same time, and with equal commitment, Ptolemy sought to predict the events that took place in society and in individual lives, for he believed that the celestial spheres were superior to the world below because they were perfect and immutable, the divine source of all heat, light, and motion.

Ptolemy's formulation for predicting future Venus positions, for example, is far more complex than either the 587-day formula from the time of King Ammizaduga or the 4- and 8-day correction schemes employed in the Mayan Venus Table. His mathematical scheme for reducing the errant motions of the planets to a component of regular orderly movement is rooted in geometry and based specifically upon the principle that the planets move on circular orbits, the centers of which move about other circular orbits around the earth, which is fixed at the center of motion. The whole system operated like a series of mechanical gears perfectly fitted together, each planet being driven along its course through space and making its place in the sky known in advance, at least to one appropriately initiated.

Ptolemy was as much an organizer and systematizer as he was a discoverer. The *Tetrabiblos* tries to express what an educated person ought to know about astrology in the style of the times. For example, he felt compelled to unite what was known to his predecessors about the function of the body and its relation to the cosmos. Greek medicine taught that, just as the universe was composed of four basic elements, so does the body consist of four basic fluids: blood, which is warm and moist; yellow bile, warm and dry; black bile, cold and dry; and phlegm, cold and wet. We can see the influence of the Pythagoreans in Ptolemy's attempt to find number in nature. These basic fluids—humors they called them—could be aligned with the seasons of the year: spring, summer, autumn, and winter, respectively, and paralleled with the ages through which we all pass: childhood, youth, maturity, and old age. Is not the quick pulsing of hot blood in the veins like spring and youth, the slow lethargy of the phlegm like winter and old age?

The planets fit into the logic of hierarchy by the assignment of ages and humors to them. Saturn, for example, being in the highest orb (in other words the one farthest from earth), would govern old age because it moved slowest of all. It would dominate the phlegm, which is the most viscous and slow moving of the body's fluids. Being farthest from the sun's warmth, Saturn would make us cold bellied—fill us with phlegm and make us cough. Sometimes the planetary associations were so complex and detailed that their meaning, at least for us, has become obscure. For example, Saturn governed the right ear, spleen, bladder, and bones; Jupiter the sense of touch, lungs, arteries, and male semen—being between cold Saturn and fiery hot Mars, he was lukewarm. Mars covered the left ear, kidneys, veins, and genitals. As his red color and nearness to the sun indicate, he emitted a parching heat. In his lower orbit he governed black bile, which is why he could make us feel melancholic. Venus warmed less than Jupiter but moistened more, for her large surface caught many earth vapors. She governed the nose, liver, and muscles while Mercury ruled the tongue and thought, bile and buttocks. Because of his high velocity, he could be a potent cause of change. The moon accounted for the sense of taste, the womb, and the belly. By proximity to the earth the moon, too, was affected by the rising vapors. This is why she made bodies soft, causing them to putrefy. Finally, the sun governed eyes, brain, and heart. He warmed and dried, and the nearer he came to our part of the world, the more heat he produced.

Metals, herbs, body parts, individual organs, types of discharge—all were compartmentalized among the humors and thus among the planets. Humorous to us (in the modern corrupted sense of the word), the theory of humors simply states that when your juices are all in balance you function well.* This means you need to be aware when things are not in proper influence so that you can act accordingly.

The basic logic connecting astrology with medicine, though hinted at in the fifth-century B.C. writings of Hippocrates seems not to have

---

* Our list of birthstones by zodiacal sign is a remnant of the principle of associating terrestrial and celestial properties and entities (mine is amethyst, once supposed to prevent drunkenness—a natural malady associated with watery signs such as Pisces).

developed fully until second-century B.C. Alexandria. The correspondence principle goes something like this: Different parts of the body are "in sympathy" with particular signs of the zodiac. When a given sign is, say, malevolently aspected by a planet, the part of your body allied with that sign can get sick. To get better you must act to make the sign stronger by administering the appropriate herb, or medicine, or amulet made of a substance that corresponds with that sign.

This doctrine comes down to us through the Hermetic Corpus, a set of works written in Greek from the time of Ptolemy and named after Hermes (a form of the earlier Egyptian god Thoth), the god allied with all knowledge associated with writing. This includes the medical corpus, astronomy, astrology, alchemy, and other disciplines. (Today our term *hermetically sealed* connotes a high degree of cleanliness or sterility, for it literally means fused or sealed by chemical means.) Horoscopic astrology associated the hierarchy of planets with the qualities of humors on the basis of a resemblance of action or property (for example, quick-pulsing hot blood and fleeting Mercury). The aspects from one domain of nature automatically took on a relationship with those from another. In these acquired likenesses there was no need of the causal connection our modern minds require.

This natural way of ordering things, by hierarchy and by alternation between extremes, just the way society functioned and real people behaved, was applied to all things, including the planets. In the Greek-Babylonian system, the intensity of the influence of a planet upon terrestrial matters depended on the magnitude of its sphere. The higher in heaven the planet rolled about the fixed earth, the more power it held. As you passed down the heavenly hierarchy, like moving down the political stepladder from high nobility to common bureaucrat, the influence became weaker. The alternation principle added the elements of good and evil to the hierarchy, switching back and forth as one moved from the highest to the lowest sphere.

Accordingly, Saturn, which lumbers along slowly and ominously on the outermost orbital lane of the zodiacal roadway, was termed the Greater Illfortune, acknowledged by universal experience to be the most potent, evil, and malignant of all the planets. On the contrary, Jupiter, Saturn's next-door neighbor in the earth-centered universe, was the most propitious of planets, the Greater Fortune, although his influence was somewhat weaker than Saturn's.

Next down the celestial ladder, Mars was inferior only to Saturn in malefic influence; he was called by the old astrologers the Lesser Illfortune. Whereas the influence of Saturn is like a fatal consumption, Mars's is more akin to a burning fever. The sun came next in power, and it had generally a good influence. Venus, the next, bore the same relation to the Greater Fortune, Jupiter, that Mars bore to Saturn. She was the Lesser Fortune, and her influence was in nearly all respects benevolent.* Mercury was considered a cold, dry, melancholy star with a weak general tendency toward the unfortunate, especially when ill aspected. The fast-moving moon came last in the planetary hierarchy. As the one nearest the earth, she was the least influential of all celestial bodies, being regarded by astrologers as a cold, moist, watery planet, variable to an extreme and, like the sun, partaking of good or evil as she was aspected favorably or the reverse. Thus the qualities of absolute good or evil seem to have diminished and blended together—in effect to have canceled each other out—as one moved down the celestial hierarchy toward earth.

Incidentally, this planetary series is responsible for one curious if remote derivative—our system of naming the days of the week. The original scheme was Babylonian. Each planet was believed to govern an hour of the day, starting with the first hour of the twenty-four-hour day. That belonged to Saturn, and after him we name the first day of the week—it used to be Saturday. Passing down the list of seven heavenly bodies three times plus three left over to make twenty-four carries us on the next round to the fourth entry in the list, the sun. He governed the first hour of the second day, called Sunday. Again skipping three we come to the name of the third day, Monday, after the moon, then Mars's day (Tiw, the Nordic equivalent of Tuesday, has survived in our culture—though the Mardi Gras has a familiar ring for us). Then Wednesday (Woden's Day) after Mercury, Thursday from Thor, the Jovian equivalent, and, finally, Friday from Fria or Venus. (We acquired the names of our weekdays through Anglo-Saxon mythology, into which certain Nordic equivalents insinuated themselves, but if we look closely we can see the ancient Roman conquerors still reverberating through the list.)

The planets even influenced the ancient college curriculum. The seven liberal arts, or skills of which all freemen in the Greek world could partake, consisted of grammar, dialectic, rhetoric, arithmetic, geometry, music, and astronomy (called astrologia!)—the same as the number of planets. Later the first three language-based studies, absolute requirements for entry into the rest, were codified as the Trivium. (Curiously, *trivia* means the exact opposite in our culture, and these subjects are often given trivial regard in today's more socially relevant college curriculum.) The more fearsome four, attained only by the gifted few, came to be known as the Quadrivium.

And let us not forget another legacy of the planetary series, the seven deadly sins: sloth (Saturn), anger (Mars), lust (Venus), avarice (Mercury), pride (Jupiter), gluttony (the sun), and envy (the moon). Each of

* The break in the alternation sequence at this point probably results from a perversion of the older Egyptian planetary series, which ran Saturn, Jupiter, Mars, Venus, Mercury, Sun, Moon.

these is the ever-present negative side of the empowerment of its planet—an all-consuming sun leads to gluttony; the moon's paleness to envy; pride stems from an excess of power in the most powerful of gods; and so on.

The signs of the zodiac constituted another celestial list, almost as familiar to nonbelievers as to practicing astrologers. What scientifically minded, rational individual has not taken an occasional peek over his or her morning orange juice at the daily astrology column next to "Dear Abby" or the comics? Unlike the planets, the signs remain fixed with respect to the stars, but they move together as the sphere of stars appears to rotate daily about the earth. In ancient Babylonia Aries the Ram was the first 30° segment of the ecliptic, measured from 0° to 30° celestial longitude. (The measurement begins at the vernal equinox, one of the points of intersection of the ecliptic—the median strip of the zodiac—and the celestial equator, or the extension onto the sky of the geographic equator of the earth.) Taurus occupies the 30° to 60° zone, Gemini 60° to 90°, Cancer 90° to 120°, Leo 120° to 150°, Virgo 150° to 180°, Libra 180° to 210°, Scorpio 210° to 240°, Sagitarrius 240° to 270°, Capricorn 270° to 300°, Aquarius 300° to 330°, and finally Pisces 330° to 360°.

Each zodiacal sign carries with it a list of properties capable of articulating the full range of qualities or phases of human mental and emotional experience. For every one of these properties an antithesis can be found somewhere in the composite list. For instance, the adjectives said to describe a person whose sun sign is Aries are vernal, dry, fiery, masculine, cardinal, equinoctal, diurnal, movable, commanding, eastern, choleric, violet, and quadrupedal!

What characteristics were assigned to the segmented stellar path? Whether a constellation is masculine or feminine seems to have mattered. This was based on the Pythagorean notion that odd numbers were male, even female. Whether they are human (like Virgo), bestial (like Leo), or a combination of the two (like Sagittarius) was another. Do they rise right side up (like Virgo) or upside down (like Taurus)? Are they day signs (in which the sun resides when the day becomes longer) or night signs (when the day's length decreases)? Are they aquatic (Cancer and Pisces) or terrene (Gemini and Sagitarrius)? Fertile (Cancer, Scorpio, and Pisces—animals that proliferate in great numbers) or unfruitful (Leo's mate bears cubs infrequently and then only in small numbers, while Virgo is barren)? And do they run (Leo and Sagittarius), stand (Gemini and Aquarius), or sit (Taurus and Libra) in the sky?

As the planets pass among the zodiacal signs, each exercises its motivating power, which, too, alternates between opposing states: good and evil as well as strong and weak. Moreover, the influences of the planets may be canceled or enhanced by their relationships to one another. For the Greeks, the interrelations were explained, as we might expect, through geometry. The most common relationships or aspects had their foundations in the liberal arts.

Today we use mathematics and geometry as convenient languages in which to express relationships that we believe to exist in the material world. But if you learned arithmetic in the Quadrivium in the old Pythagorean School, your teachers would have been less concerned with the processes of adding and subtracting. Instead they would have been teaching you about the real meaning of the numbers themselves— the way I spoke about them in Chapter 3—and the relationships among them, especially in terms of geometry. For example, the number 2 taken by itself was feminine. It did not matter whether you counted two roses or male twins. The number 8 was "the Cube" or $2 \times 2 \times 2$, whereas 9 was defined as "the Square" (3×3). Besides number and alternation the zodiacal signs were also related to one another through geometry. If you connect every third constellation (Aries, Leo, and Sagittarius; Taurus, Virgo, and Capricorn; Gemini, Libra, and Aquarius; Cancer, Scorpio, and Pisces), you end up with four equilateral triangles called the Trines. You can also make three squares by tying together Aries, Cancer, Libra, and Capricorn, and so on, as well as a pair of hexagons (Aries, Gemini, etc.; and Taurus, Cancer, etc.), one of which holds the masculine and the other the feminine signs. Moreover, it was the Squares themselves that forced the constellations 90° apart on the sky circle into a generally malevolent relationship. Conjunction and opposition, terms still in astronomical usage, mean signs lined up with (0°) or opposite to (180°) each other. Influence among planetary aspects alternated between beneficial and harmful as one proceeded through the list: 0°–30°–60°–90° and so on.

But of all the places a planet could be, its position in the local environment mattered most of all. Astrologers of old divided the ecliptic in yet another way, which depended on latitude and longitude, time of year, and time of day as we would express it today. To understand this local system, imagine slicing the sky into twelve 30° segments, like the sections of an orange. Make the cuts along lines symmetric about the meridian, which passes overhead and through the north and south points of the local horizon. This divides the sky into the twelve zones or houses, each of which includes a 30° strip of the ecliptic. The intersection points of house boundaries and ecliptic are the familiar *cusps* of astrological lingo. These divisions do not necessarily coincide with the division of the zodiac into signs, which are based on a different beginning point.

We can think of the house system as a set of allotments. In this arrangement, the first house is the first 30° segment beneath the eastern horizon (called the ascendant), and it contains the celestial objects that are just about to cross the local skyline and come into view. Called the House of the Ascendant, it is most influential of all, and will affect your appearance, your self, and your beginnings. Planets residing there at the time of your birth will have the most potent outcome on your life and destiny. The second house, the House of Riches, is the 30° to 60° zone below the eastern horizon. Kindred and short journeys in the third house combine foreordained knowledge of contracts and communications; inheritances reside in the fourth, children in the fifth, health as well as your service and employment tendencies in the sixth. The House of Love and Marriage is seventh, Death the eighth. To the ninth house belong long journeys and foreign affairs. Tenth is the House of Honor, and it affects your public standing and inclinations in professional life; eleventh is the House of Friends, and twelfth that of enemies.

Of course, there is no reason to think the ancients would classify character traits as we do today. For example, sexual performance in some systems is dealt with in the House of Death (the French still call orgasm *le petit mort* or "little death"); theater and merrymaking belong in the House of Children, and parenting in the House of Honor. In the houses, then, you could find information on the inclination of your state of health, wealth, wisdom, and so on.

Why divide the ecliptic into segments marked off relative to the local horizon and why assign prominence to the first segment below the eastern horizon? We do not really know, but the idea likely had a practical basis; for example, early agricultural people were well aware that the sun's rays had different effects on their crops at different times of the day. How many gardeners still insist that their plants must bask in the morning sun? And recall that worshipers in many lands were required to face the sun's first rays, which appear at dawn in the east.

Essentially, the sun, moon, and planets move through the twelve houses in a day, thus duplicating the annual solar motion, except at a more rapid pace. It is probably this parallel between macro- and microtime that led the Greeks to reason that if position on a twelve-part zodiac had influence, the same would hold true regarding position in the twelvefold house divisions. A chart that shows the positions of heavenly bodies in both the zodiacal signs and houses is technically called a horoscope. To cast a horoscope, the astrologer needs to determine the positions of the planets, sun, and moon with respect to both the zodiacal signs and the various houses. Precisely where the wanderers lay in each framework at the time and place of your birth signals possible character and personality traits, sympathetic and antipathetic influences.

Far too extensive to list here, there are many other ways astrologers of the classical world subdivided the sky.* Suffice it to say that their

---

* For example, as with the planets there were also bodily allotments. From top to bottom of both lists: Aries governed the head, Taurus the neck, Gemini the arms; Cancer the breast, Leo the sides, Virgo the belly, Libra the loins; Scorpio the groin, Sagittarius the thighs, Capricorn the knees; Aquarius the shanks and, finally, Pisces the feet.

basic guideline for doling out fatalistic allotments was the alternation between extreme human qualities as indicated by number and geometrical figure, each carrying meaning independent of the real world as we know it.

The horoscopic system was very complex, to say the least, but to give a simple idea of how it worked, let me describe the nuts and bolts of the way an astrologer puts together a nativity horoscope today. The process is not altogether different from the way the Greeks did it, although interpretations have become modernized. In the modern practice of astrology, planetary positions relative to the stars are easy to determine from modern astronomical tables, such as the *American Ephemeris and Nautical Almanac*, published annually by the U.S. government. From it, the astrologer can extract the equatorial or sky-fixed coordinates of each of the primary celestial bodies. Or a computer can supply them and mathematically convert them to the lococentric framework defined by the exact place and time of your birth. Once these corrections have been made, the zodiacal sign and house associated with each can be determined.

To give an example, at the time of my birth, Mars had a celestial longitude of 233°; in other words, its position lay 23° (counting in multiples of 30°) into the seventh sign, which is Scorpio. The astrologer would write ♂ 23° ♏. The well-known symbol preceding the number is the Mars symbol, and the one following stands for Scorpio. While Mars lay in Scorpio as seen by everyone on earth, in New Haven, Connecticut, where I was born (at six in the morning), it resided 165° below the eastern horizon reckoned along the ecliptic. This put it squarely in the middle of my sixth house. Therefore an astrologer reading my horoscope would suffix the symbol VI to the previous shorthand notation. Having marked off the locations of the signs and houses of each planet, the astrologer would then consult an interpretive set of tables on properties of signs, houses, planets, and aspects for the basic information necessary to begin to compile my nativity horoscopic chart.

How should I interpret my own ♂ 23° ♏ VI? Not being an astrologer, I cannot say, but I would presume it sheds light on my job or perhaps my tendency toward a good or bad state of health (remember the sixth house is the House of Sickness). Mars rulership would naturally connote a mildly malefic influence. That Gemini constitutes a more fortunate track of the ecliptic helps me, but its influence over the lower portion of my body might indicate that when it lay over my cradle, it correctly portended the mild hereditary dose of osteoarthritis I now experience in middle age in my knees and ankles—or it is just a coincidence, I wonder?

By contrast, powerful and beneficent Jupiter lay in my House of Riches, which I should think would have set me up for a lifetime of ease regarding money and material possessions. This, alas, has not come to pass. But, then, perhaps I need to interpret the term *riches*, as some astrologers do, to mean that I will be rich in the liberty to pursue material gains—whether I will actually acquire the fruits of my labor is quite another story.

In practice there is much more to a nativity horoscope than I have implied. My lists of properties and magnitudes of influence give only a general idea of how astrology works. They demonstrate that it *is* a system with a set of rules, though subject to far more subjective interpretation than modern scientific methodology would allow. Experienced astrologers will tell you that it takes years of practice before one can extract all possible information in detail from a horoscopic chart once it has been constructed. For example, certain combinations of signs supposedly give more details than others about future events; other combinations produce nullifying effects.

Today only a devout minority consult the horoscopic listings with any thoroughness, but it was a different story in the Latin Middle Ages and in ancient China and Mexico, where the noble classes carried out the practice on a daily basis with as much attentiveness as the common person today who seeks psychoanalysis. Although precious little detail survives, we have good reason to believe that ordinary people then, too, had shamans and diviners with whom to consult on the street corner or in the plaza about what the stars might foretell. Among remote contemporary people of highland Guatemala, for example, the Mayan daykeeper still practices his craft. They say you need to be born on the correct day of the 260-day calendar to become a daykeeper, for only then can you receive the "lightning"—a kind of soul that moves in your blood and endows you with the ability to receive word from the supernatural world. Like our therapists, these modern astrologers service the sick and the troubled, those about to marry and those who have committed adultery—anyone concerned about the future, anyone who has a burning question.

MANIFOLD ZODIACS

I once offered a description of the Mayan zodiac to a friend: Its sky serpent's body was segmented into the Peccary constellation, the Deer, Sea Monster and Vulture, Tortoise and Rattlesnake—even a Scorpion like our own. "You mean other cultures had astrology, too?" he asked in amazement. That other cultures of the world practiced astrology often surprises popular readers. I have briefly charted out the historical course of the Greco-Babylonian system of astrology because it is the one, in modified form, that clashed with the new science and technology of the Renaissance, which, as you will see in the next chapter, reimaged the planets in a totally different light. But lest I convey the erroneous impression that only the ancestors of the West lived their lives by the stars, let me mention a couple of other places—China and India—where a similar tradition of planet watching developed, most of it quite independent of the Western world, I think.

Time reckoning was a preoccupation with the Chinese. In their highly bureaucratic state, they needed to keep tabs on a host of religious festivals. They faced the dilemma of reconciling the highly visible and graphic short-term advantages of lunar timekeeping with the slower, seasonally tuned pulsebeat of the solar year—the same dilemma encountered by the Maya and the Babylonians (outlined in Chapter 4). Early Chinese astronomers kept close watch on the position of both sun and moon against the background of distant stars. For example, they reckoned the first day of summer by the last sighting of Antares (in our constellation of Scorpio) in the southwest, just after sunset. Our astronomers would have said: Sun in Libra, the next constellation below the horizon. With each marked astronomical event came an omen. Much as in Babylonian astrology, statements accompanying observations pertained to emperor and empire, for, like the rulers of Maya Copán and Palenque, the Chinese emperor was an extension of celestial forces—the Son of Heaven. Many other constellations even bore imperial terminology; the Purple Palace (consisting of part of our Little Dipper) was composed of stars named Emperor, Empress, and Royal Prince.

Fourth-century B.C. Chinese astronomers followed the moon from station to station seeking symmetry between its behavior and that of the emperor: "When a wise prince occupies the throne the moon follows the right way"; "When the high officials let their interests prevail over the public interest, the moon goes astray toward north or south"; or "When the moon is slow, it is because the prince is rash in punishing."[11] Although there were twelve stations for the sun, the Chinese found it more bureaucratically suitable to lunar forecasting to devise twenty-eight moon stations, one for each day of the month. Han dynasty astronomers knew of the 8:5 ratio of seasonal years to Venus periods as well as of the four divisions of the Venus cycle. However, they listened more closely to the presages of Jupiter and Saturn and, to a somewhat lesser degree, Mars. Every time those planets entered their retrograde loops their actions warned the state; thus, "When Mars retrogrades in Ying-she, ministers conspire and soldiers revolt."[12] And all of this took place long before any appreciable contact with the West.

Further west from remote China it is difficult to disentangle indigenous astrology from those versions shaded with Babylonian influence. In Indian epic tales such as the *Ramayana* and the *Mahabharata*, traceable to the first few centuries B.C., the planets were said to influence the world, especially when they moved in retrograde or lay in close conjunction with one another in certain constellations. Hindu time reckoning and its relation to astrological prediction has a foundation that would sound familiar to a Mayan priest. It is based on time cycles of various duration that fit neatly together. Biggest is the *kalpa* of 4,320,000,000 years, 72,000 of which make up the life of Brahma or the duration of the universe. Somewhat like the Mayan Long Count, the kalpa is subdivided, by 1000, and scaled down further into more manageable time periods called *yugas* or ages. The beginning of the briefest of these cycles was geared to an event computed by the astronomers actually to have taken place in the sky. The event was a conjunction of all the planets in the constellation we call Aries at midnight of February 17–18, 3102 B.C. (coincidentally not very far from the Mayan creation cycle, which began on August 12, 3113 B.C.).

The object of the chronological game in ancient China and India was to link real time—durations that are observable and sensible to us—with the time ordained by the planetary deities. Only then could each deity's messages be read in the stars. Later Persian astrologers—strongly influenced by India—devised a clever way to do this. Saturn and Jupiter conjunct every twenty years. For Persians, human history was written in the unfolding of influences of Jupiter-Saturn conjunctions. In 240 years, a dozen conjunctions cover the signs of one triplicity. A triplicity consists of one-quarter of the zodiac, or three constellations located 120° apart at the points of an imaginary triangle. This division was considered to be related to the four elements and the qualities they transmit to earth via the stars. Thus, Aries, Leo, and Sagittarius are fiery; Taurus, Virgo, and Capricorn earthy; Gemini, Libra, and Aquarius airy; and Cancer, Scorpio, and Pisces watery. An astrologer dating the ancient flood would calculate when major conjunction cycles took place in watery signs of Jupiter and Saturn. Not until 960 years (and 48 conjunctions) have passed will the planets have encountered one another in each zodiacal sign of each triplicity. Shifts between triplicities attended dynastic alteration, but those on the order of our millennia meant even more drastic change—natural disasters such as earthquakes, monsoons, and tidal waves.

We know Western tradition has it that all the planetary wanderers as well as the sun and moon move on the central strip of the zodiacal band of constellations. The Sumerians called it "the way of Anu," a road lit by bright stars, counselor gods or consultants who surveyed what happened in their own zones and advised the gods about future undertakings. The northern region of the zodiac near the Tropic of Cancer constituted the way of Enlil or Bel, Lord of the Earth, whereas the southern part, the region of the Tropic of Capricorn, was the way of Ea, god of the waters.

But doesn't it seem curious that zodiacs on both sides of the globe are divided into zones delineated by star-to-star constellations? The Old World skyway is populated by a man holding a jar of water, a virgin, and a set of twins; its bestiary includes a lion, bull, ram, and scorpion. There are even imaginary composites such as a half-goat-half-fish creature and a centaur—half man, half horse. The Mayan zodiac of the New World, as we have seen, is populated with similar creatures. And the solar zodiacal twelve of the Chinese includes ordinary dog and rat as well as exotic dragon.

Was there once a single culture that spawned and propagated the idea of peopling the sky with imaginary denizens? Tempting as the idea sounds, I think not. I find it easier to believe that we invent the cosmos after ourselves. The desire to construct a zodiac is motivated by the natural tendency we all have to seek the familiar in the unfamiliar, to express the unknown in terms of what we know, in this case what we all can plainly see in the sky. We lend worldly attributes to the unseen forces of nature. I think it likely that pastoral people all over the world who watched for the signs of the seasons—where the sun resides during each of the twelve lunar months over the course of the year—would likely assign names to the sun's place on the celestial roadway; names that are associated with the natural attributes of those seasons. Vestigial traces of this idea can still be found in our modern zodiac, even though it has been changed time and again. For example, all the signs that have watery-sounding names—Capricorn, the sea goat; Aquarius, the water bearer, Pisces, the fish—hang together and consecutively in the sky as a reminder that it was during the rainy season that the sun once passed through their houses. And some of the Hebraic months, such as Nisan (Sacrifice) and Airu (Blossom), still reflect traces of activities that took place in the local civic, ritual, and agricultural yearly round.

CHANGING THE IMAGES
OF THE PLANETS

What force, we ask, can affect the bonds of sympathy between ourselves and the stars? What is the essence of this power? Does it come, as the Maya believed, from the stabbing rays that penetrate us or from certain virtues and emanations that we breathe or rub up against, as some of the old Babylonian inscriptions say? The appropriateness of the explanation would depend on where and when it was offered in human history—as well as to whom and for what purpose.

Is the force that moves all things indifferent to and unaffected by the object it influences? Aristotle would have said yes. For him all objects in the universe move toward a final resting place; the final cause is remote to the process of motion, disconnected and aloof from it, a teleological forward pull from the great beyond. In Aristotle's world the influence of one body on another was not mutual, as Newton later had it. For the Greeks, objects behaved more like people.

Striving and yearning, like love, hope, and despair, are human qualities, but, despite Greek geometrical logic, they remain embedded in the vocabulary used to describe a universe whose animate qualities persisted in our minds in one or another form up until relatively recent times. If Newton had written his universal law of gravitation four hundred years earlier, it might have read: "Every object in the universe is mutually sympathetic or antipathetic to every other; through influences that are comprehensible only to the gifted through revelation rather than to all through the intellect."

The key word here is *influence*, that ubiquitous term used to describe the relationship between planets and people over the ages.[13] Originally it meant a flowing in or influx of power, or a substance that endows power. When we think of gravity, we imagine an empty space populated by bodies that interact without ever coming into contact with one another. We say that masses influence one another through the principle of action at a distance. For the ancient Greeks who devised astrology, the planets were not moved by gravitation. Rather, the wanderers were self-willed; like people they had soul to endow them with life. Medieval minds conceived the power that connected all nature's components as the *pneuma*, an elastic, airlike, invisible fluid that permeated everything and endowed the entire universe with a collective soullike quality; it was literally the "breath of the universe" (our modern term *pneumatic*, applied anticlimactically to tires and drills, is about all that survives of this concept). Still later, influence was believed to travel through an all-pervading ether. Physicists as late as the twentieth century still searched for it, even though they no longer regarded the substance as an animated, breathing, divine will.

Some readers may be disturbed by my descriptions of how astrology worked for true believers of a bygone age. They may ask: How can one trust one's destiny to the intellect of a supersensitive priest, someone removed from the concerns of daily life? What if people in high places practiced the art today? Think of what could happen to the world! (Of course, Americans are well aware that the wife of a two-term president in the eighties regularly consulted with an astrologer.)

For most of us, however, astrology's tenets seem too illogical. We wonder, why should the planets' power of influence *increase* with distance when the principle of gravitation, which can be documented through physical experiments, works exactly the opposite way? Why isn't the sun the most powerful influence if the magnitude of a celestial object determines its power? Modern scientists have discredited astrology on the ground that it has failed to synthesize its theories into a comprehensive whole that can explain the movements of all the planets around the sun. We criticize ancient astrologers for not checking their observations against existing theories or not seeking to go beyond magical appearances to try to understand the mechanism that underlies all celestial motion.[14] Although we seem quick to praise the old Babylonians for their legacy of orderly notebooks filled with carefully acquired celestial observations, we accord little value to their cosmogonic ideas or the relationship of their beliefs about the sky to religious practice and daily life. Instead, we complain, Why weren't they like us?

Much of this railing against ancient planetary mysticism is an extension of modern science's war on contemporary horoscopic astrology, which has little to do with the worldview of its true ancient ancestor. Today's astronomers have expended considerable effort debunking the old art. It has even been formally denounced by decree at organized sessions of professional astronomical meetings. These are just a few of the objections to the logic of the modern astrologer:

1. Because we understand birth as the culmination of a developmental process that begins with the union of sperm and egg, why is the moment of birth, rather than that of conception, employed by astrologers in making their predictions? (This is the same issue that plagues the abortion controversy.)

2. How can astrologers have cast correct horoscopes before the distant and fainter planets—Uranus, Neptune, and Pluto—were discovered? The "magnitudes of their spheres" are all greater even than those of Saturn, the Greater Illfortune. Imagine what our ancestors must have overlooked in their ignorance of the power inherent in Uranus, Neptune, and Pluto and all the other unknown planets that still lie in the outer realm of the solar system awaiting discovery!

3. Why do modern astrologers stubbornly insist on using the placement of the signs along the ecliptic as they were situated three thousand years ago in ancient Babylon (see definition on page 146) when we know full well that the precession of the equinoxes, the long-term wobbling of the earth's celestial pole among the constellations, has moved the starting point of the zodiac from Aries to Pisces, thereby changing everyone's sun sign?

But most of these objections are based largely upon knowledge that we value in our scientific way of understanding the universe. As we know it, the solar system is comprehensible only in terms of a sun-centered model. Human life can originate only as the result of a single conception-birth process, and interactions among the planets can happen only via gravitational attractions that obey Newton's inverse-square law.

But the first objection doesn't consider the ancient Greeks' beliefs. In their world, birth was the culminating event of the prelife cycle. The fetus inside the womb only gradually took on its identity, first acquiring seed, then pneuma (breath). It drew blood and thus acquired flesh. In the last stage the fetus was nourished by mother's milk rather than menstrual blood. It was not whole until all processes were complete. Just as life ends at death when one leaves the world so, in the Greek view, it did not begin until birth.[15]

The second and third objections are based upon gravity and cause and effect, both mainstays of modern scientific thought that have nothing to do with the foundations of astrology; therefore modern demonstrations that these do not bear out the ancient methodology have little value—except to teach us that we ought not believe in something we gave up long ago.

Both modern astrologers and their detractors need to be reminded that *ancient* astrology's message was intended to be delivered to the doorstep of a different house. The three I have listed are certainly valid objections to the way *today's* astrology is practiced because the assumptions they cite lie very far from current consensus about how the universe behaves. But the wrath of the debunker ought not be transferred to the archaic practice of astrology, which sought and acquired different forms of knowledge, accorded importance to certain categories of information that remain unemphasized in our astronomy, and integrated this knowledge with deeply held beliefs concerning the relationship between human beings and nature.

Debunking astrology is no innovation. St. Augustine was an outspoken critic in the fourth century A.D., long before Christianity absorbed the pagan planetary deities and refocused their images. In his *Confessions* he tells the story of a friend who was born at exactly the same time and in the same household as the child of his family's servant. Their horoscopes were cast by two different astrologers; needless to say, because their stars were the same, the horoscopes were identical. Yet Augustine's friend attained great wealth and honor while the slave spent his entire life in servitude. Augustine tells us that he turned his back on astrologers and their illusory claims to predict the future, along with their insane ritual. Like the modern scientist, he rejected astrology because its outcome conflicted with the value structure he believed in and out of which he acted.

Augustine also tells the story of a man so dependent on the stars that he calculated the hour to conceive a child with his wife, so that the planets would be auspiciously aligned at the time of birth. But when his wife came to term a bit earlier than planned, he stood at her bedside pleading, "Don't let the child come forth yet, for if he is born now he will be ugly and unsuccessful."[16]

We need to trace the pathways of planetary understanding from the Classical world to essentially the European Renaissance in order to understand how we acquired our present, decidedly nonastrological, view of the planets. In much of the world Christianity has demonstrated a remarkable capacity to syncretize pagan religions, to appropriate their imagery and transform it to reflect the most readily definable Christian value or concept. This is one reason for its remarkable success. For example, in Mexico at the time of the Spanish Conquest, Quetzalcoatl was seen to possess many of the qualities of St. Thomas, so the adopted St. Thomas in Mexico took on many of the ancient feathered-serpent god's characteristics. And who could be a more ideal sun god than Jesus Christ? When he moves to the southwestern corner of the world in the dead of winter, he is crucified and with him all life ebbs; but, after passing through the underworld, he is resurrected and his returning warmth and light restores order to the universe, bringing rain to fertilize the new maize. The cross upon which he was crucified becomes the Ceiba of Mayan creation mythology, the first world tree, the cross-arms made of a pair of upturned horizontal branches.

One of the triumphs of modern church scholarship is the discovery of the persistence into the Middle Ages of planetary imagery from pagan Babylonia. The identification in early Christian and Islamic manuscripts of the same personifications of each planet and the slow transformation of ritual imagery through time is testimony to the holding power of astrology.

As in the pagan Old World, ecclesiasts coopted the characteristics of old Babylonian Mercury (Figure 3-1e) and transformed him into a pious and scholarly scribe (Figure 3-1g), one who had his origins in the Chaldean writer god, Nebo. Jupiter, the judge, was an altered image of Marduk of Babylon. Figure 3-1h shows him as a monk with chalice in one hand and Christian cross in the other, looking very much like any other figure of that age. And, as you will see, aspects of Venus became engaged in a complex dialogue about the nature of true Christian love that lasted well into the Renaissance. Using the familiar to develop the novel and less familiar philosophy made it an easier "sell."

This radical new Christian philosophy arose in the stable period of the Roman Empire (about the first century A.D.) but not without serious competition. An old planetary cult known as Mithraism had filtered in from the Middle East and begun to take root in the empire's upper classes. It was based on the teachings of the Persian philosopher Zoroaster, and it centered on the Persian sun god Mithras, whose dualistic counterpart was Saturn, the sun of the night. Because he was the supreme ruler of time, Mithras's wanderings about the zodiac were duly noted.

This ancient cult is especially noteworthy for two reasons: first because it relied heavily on astronomical concepts, and second because at one time it was the only serious rival of Christianity. Its planetary fatalism appealed to soldier, bureaucrat, and merchant alike, maintainers of the status quo in Rome's social sphere (women were excluded from the cult). It succeeded precisely because it appealed to everyone— at least everyone who had power and identity in that vast empire. Yet its believers worshiped in underground temples. Secret knowledge is sacred knowledge, and, like members of our modern secret societies, fraternities, and sororities, Mithraists took vows of silence in elaborate initiation ceremonies and swore never to reveal their mysteries.

The Mithraic temple iconography is notable for its stark and direct visual symbolism and in particular for the order accorded the planetary positions. For example, the Mithraeum at Ostia Antica near Rome known as Sette Sfere (Figure 5-2) (named after the seven spheres or seven gates through which Ishtar passed when she descended into the underworld), is a virtual map of the cosmos.

The zodiacal signs are laid in counterclockwise order around the tops of a pair of cardinally aligned benches. If we interpret them as sun signs, the year of the seasons begins in the east and moves northward toward the west, just as the sun does in summer; finally the flow of the annual cycle of time returns via the south to the equinox sunrise in the east. There worshipers confront Mithras, surrounded by solar and lunar torch bearers, along with an odd assortment of animals—a raven, snake, lion, and scorpion among them—in a bull-slaying scene called the tauroctomy. The planetary signs gird the sides of the benches, but their order—Jupiter-Mercury-Moon-Mars-Venus-Saturn-Sun—differs from the Greco-Roman ordering. The hierarchy makes no sense, unless, as classical historian R. L. Gordon argues, you look at it as the order of planets the way they were perceived in the sky on the night of creation. In fact, the order is the same as that prescribed in astral lore and found on reliefs, for example, at Sidon (today in Lebanon).

The inside of the Mithraeum depicts the horoscope of the birth of the world, which occurred the night Mithras killed the bull. Such realism is in stark contrast to the more opaque planetary imagery that soon became identified with the emergence of the New Israel. Mithras was the lord of time and a sun god in ancient Persia, and these roles help explain why he is so often draped with zodiacal figures. But what would a god of time accomplish by killing a bull?

FIGURE 5-2. PLANETARY SYMBOLISM IN THE TEMPLE OF A SECRET SUBTERRANEAN CULT OF THE LATE ROMAN EMPIRE

The Mithraeum of Sette Sfere at Ostia Antica, near Rome (ca. first century A.D.) shows symbols carved on a bench in the order in which they allegedly stood on the night of the creation of the world. (R. Gordon)

Historian David Ulansey has recently offered a decoding of the tauroctony scene.[17] Apparently the animals represent an actual ordering of constellations based on the sun's position in the constellation of Taurus at the spring equinox. Now because of the precession of the equinoxes, the place where the sun rises on the first day of spring would shift slowly with respect to the zodiac, moving from one sign to the next in about two thousand years. It lay in Taurus in the third millennium B.C., moved to Aries by the first, then to Pisces, where it lies today. What we attribute to a mechanical precession our predecessors would have ascribed to some sort of cosmic deity empowered to alter the star-fixed ages. According to Ulansey, the slaying of the bull by Mithras depicts the central mystery of the cult, the power of their god alone to move the entire universe, for by slaying Taurus the Bull, Mithras allowed the sun at the equinox to move into the next house of the zodiac. We might say that Mithraists used their knowledge of the precession of the equinoxes to devise a religious scheme that overcame the sources of fatalism. This planetary mystery cult derived from the same area at the same time as early Christianity, and it responded to many of the same needs. Had the emperor Constantine opted for Mithraism instead of Christianity in the fourth century, the world today might well be different. This is why Ulansey terms Mithraism "the road not taken."[18] But once the empire and its army vanished, Mithraism died, and Christianity was left alone in the field.

The early Christian opposition to astrology was not because there was anything inherently wrong with it, but simply because the Greek philosophers who practiced it were pagans. If astrology was part of pagan wisdom, then it must oppose God's will. How could God *and* the stars rule human destiny? Nevertheless there were aspects of astrological doctrine, such as truth by revelation and the dialogue between priest and client, that appealed to certain early sects. Compromises were made: In medical astrology the stars ruled the body while God guided the soul. The Christians were quick to appropriate many of astrology's supportive aspects, along with the attending celestial imagery, where it suited them.

"Astrology now is the science of the stars of Christ," wrote Tertullian, an early Christian philosopher.[19] As believers were weaned away from pagan polytheism, planets still operated as ominous signs; they were just changed into intermediators who announced the intent of God. They were never allowed to act contrary to his will, although humanity's free will, with the grace of God, could act to overcome stellar influence. The celestial world, populated by both the angels of God and the sidereal demons of Lucifer, became a battleground on which to confront questions about the nature of sin, the qualities necessary to lead the good life, and, above all, the nature of human free will.

In the Middle Ages, the planets spoke to our European ancestors through both God's word and hidden natural laws—call it natural theology. The language of these laws consisted of the number symbolism and geometry of the Greeks combined with an even richer mathematical legacy endowed to Europe by the Arab world. As we judge it with hindsight, this science is a muddled mixture of alchemy and magic. Astrology revolved about the periphery of such proto-scientific concerns, but within many of its observations lay rational principles and ideas that we might regard as at least approaching correct scientific explanations of natural phenomena. For example, an ingenious medieval philosopher once explained the tides by the production of vapors through the interaction of moonlight in the depth of the ocean. The differential upward pressure of such vapors caused tides to rise and fall when the moon stood overhead. Just as the noonday sun becomes hotter, the moon could create more virtue and consequently a higher tide, and while it lay nearer the horizon (at low water) its obliquity did not permit it to be as effective. Although this model fails to explain why there are two high and low tides a day, it possesses all the inherent qualities of mechanistic explanations passed down to us by the Greeks that are still a vital part of science. As it did to our Middle Ages predecessors, Greek science appeals to us because we revere it as our legacy.

In stark contrast, at about this same time a major conjunction of planets in Libra was interpreted by astrologers to portend a great windstorm. Panicky people throughout the countryside commenced to dig holes in the ground for protection—the first civil defense shelters. The big event was slated to take place in the year 1186, and, rare as it was, it would also be linked to a pair of eclipses. As they have always done with celestial events, from total eclipses to the returns of Halley's comet, the experts weighed in with their interpretations about it well in advance. Practically every astrologer in Europe made a prediction. The Franks would benefit at the expense of the Saracens, a Sicilian astrologer anticipated. A new Christian prince would arise, claimed an Anglo-Saxon. An English sky priest went into a trance, babbled thirty-three predictions in the form of poetic couplets in bad Latin, and then dropped dead; a Spaniard calmly announced that all the good and evil influences would cancel out and produce no effect whatever. He seems to have been closest to correct.

The medieval courts of Europe were overrun with astrologers; they were consulted by bishop and king, priest and prince who wanted to know what would take place in heaven and especially when it would happen. One of those who was prominently involved in prefiguring planetary conjunctions was Cecco d'Ascoli, astrologer of the court of Florence, a member of the Franciscan Order and special adviser to Florentine medical doctors. "A doctor must of necessity know and take into account the nature of the stars and their conjunctions," he tells his followers in his *Astrological Principles*.[20] Then he goes on to list all the plants and herbs associated with each planet so that they might be administered at the proper time. Sadly, power and fame led Cecco to transcend the limits of his discipline. When he began to dabble in astrological predictions based on the birth of Christ, the coming of the Antichrist, and the end of the world, he landed in front of the inquisitor. In 1327 he was burned at the stake, whether for astrological malpractice or because of political intrigue we cannot say. One of his judges, the bishop of the city of Aversa (also a Franciscan), regarded Cecco as a supporter of the rival city of Cesena, which in turn supported the breakaway faction of Franciscans to which Cecco belonged.

Despite its incipient ability to seek rational explanations for physical phenomena, we cannot understate how closely the medieval mind connected worldly with celestial affairs. In 1348 the University of Paris Medical department reported that at one o'clock in the afternoon of March 20, 1345, a conjunction of Mars, Jupiter, and Saturn in Aquarius acted to produce one of the world's most monstrous occurrences.

Jupiter—warm and humid by nature—drew evil vapors out of the earth. Then hot and dry Mars ignited them, while evil Saturn spread them in the form of a great infection. The Black Death killed off nearly half the population of Europe over the next three years. Today we have traded in the star-crossed planets for a deforested Amazon, radon gas, and a hole in the ozone layer, but we are no less conscious about the effect on our future well-being of the environment—an environment over which the medieval astrologer had far less control than we.

Most astrologers of medieval and early Renaissance Europe believed that the causes that underlay the behavior of nature were impervious to detection by the senses and the purview of the intellect. Those of us who are in the habit of acquiring scientific truth by observation would term these causes occult. Nor could reason account for them. Only those to whom the principles were revealed could know and experience these mysterious forces, or at least acquire a more intimate understanding of them. Hence, our word *mystic* has come to mean an individual who, by gift or special status, perceives and understands whatever is hidden or cannot be reasoned. This concept of a spiritually elite class of shaman endowed with extraordinary or supra-rational powers also permeated New World social systems. As you have seen, in many hierarchically structured societies, it resulted in a philosophy of nature in which the people at the top—king and priests alone—possessed the power to understand the forces of nature.

Such systems of belief seem unsuited to our egalitarian world, whose physical and biological workings we feel ought to be accessible to everyone, at least anyone who cares to exercise the necessary and often complex process of quantitative reasoning coupled with the careful and precise use of the senses in order to arrive at scientific truth. Those who acquired the power to control such a valuable font of knowledge would necessarily exercise that grasp of the occult in secrecy, for it harbored the seeds that could alter the course of society. We can legitimately use the words *mystical* and *occult* to describe the perspective accorded the planets during this long period between Rome and the Renaissance when the Christian religion dominated Europe. In medieval times you have seen that the pagan intellectual qualities once conferred upon the planets were gradually altered and transformed to better represent a heaven and earth ruled by Christian principles. Philosophers and scientists had slowly shifted their stance on how to express nature's wonders. Now they were less concerned about experiencing the natural world directly and more interested in mythologizing it abstractly.

Allegory means the representation of one image disguised within another, and the process of transformation of the planetary gods into allegorical figures is fascinating to pursue, so imaginatively did late medieval and early Renaissance writers and artists develop and express planetary symbols. If we observe carefully enough, we can penetrate the heavy veil of pen and brush in a new age that reinvented the cosmos and rediscovered the underlying planetary imagery of their pagan predecessors. What, for example, is so scholarly about the image of Mercury? The reasoning differed little from that of the ancient Chaldeans. Swiftest of all the planets, Mercury would be the logical one to mediate between gods and mortals, for he skims low over the clouds, continuously probing the intellect. When he appears, he seems to remove the low-lying clouds that dissipate after the storm. This godly function is an excellent parallel for one whose skill lies in exposing the soul, for dispelling the clouds that veil the mind. Thus, the planet Mercury became the Christian "supreme watcher."

But Christian Mercury, like Venus, has a dual aspect. First, he turns away from the world in detachment, only to return as Zephyr, the wind god. Then he blows away the clouds, enabling us to penetrate closer to the soul, which, in the old Platonic tradition, harbors real truth and always lies out of reach even of the highest human intellect: "The soul is above the realm of the Moon, of Mercury and Venus, of the Sun, of Mars, of Jupiter, of Saturn, and of all the signs which are in them: it is above the 72 constellations," said St. Bernard.[21]

What is beauty? What is love? Renaissance philosophers were not preaching morals. Refreshed by a reacquaintance with the texts of antiquity, they had begun to embark on an inquiry into the essence of being human, and old, transformed images of the planets figured prominently in their dialogue. Recall the Babylonian Ishtar—the Venus of love and war. Her voluptuous, ethereal side is represented by the evening star that hovers over the setting sun, while her carnal nature reveals itself in the love goddess's morning star aspect. Renaissance philosophers recast celestial Venus in these two aspects in a rather complex theory of love. One is contemplative, divine, and immaterial, the kind we still call platonic and to which only prophets and saints can aspire. The other Venus is tangible and terrestrial; she tempts the male imagination through visual, then tactile sensation, ultimately into debauchery.

Marsilio Ficino and Pico della Mirandola were among the philosophers of the fifteenth century who produced rather lengthy tracts on the nature of love. They classified and subclassified types of Venuses, the parts of the soul to which each corresponded, and the reaction of the male libido to each form. They even produced a manual on the metaphysics of kissing. According to Ficino, debauchery was a disease, a kind of insanity ultimately traceable to an imbalance of humors in the heart. Like the self-awareness manuals that dot today's best-seller lists, these dialogues on the nature of love influenced every courtier of the refined and enlightened societies of Renaissance Italy. All educated citizens made it their business to know the many kinds of love and how

FIGURE 5-3. PLANETARY INFLUENCES WERE DEPICTED ALLEGOR-ICALLY DURING THE RENAISSANCE

5-3a. (*above*)Titian's *Sacred and Profane Love* [ca. 1515]
(Borghese Gallery, Rome)

5-3b. (*opposite page*) Botticelli's *Primavera* [1478]
(Uffizi Gallery, Florence)

5-3c. (*below*) Botticelli's *Birth of Venus* [ca. 1485]
(Uffizi Gallery, Florence)

they might be sought and expressed in painting and in poetry, in literature and in sculpture.

Not many of us who view the Renaissance paintings in Florence's museums or in Venice's churches are aware either of the wealth of planetary imagery deeply embedded in their composition or of the possibility that each piece belongs to a sequence that expresses a story with celestial roots. According to Renaissance art historian Erwin Panofsky, the twin Venuses in Titian's *Sacred and Profane Love* (Figure 5-3a), painted about 1515, typify the two forms of love Ficino and Pico described. Nude Venus to one side is the universal, nonmaterial yet intelligible form of beauty. She carries an eternal flame and beckons to her fully clad sisterly image as if to share the place where she is seated. Beauty in its tangible form is represented by the clothed Venus, who is flanked by animals connoting love in its carnal and more fertile terrestrial form. Cupid, the intermediary between heaven and earth, lies between them, stirring up the waters of a cosmic fountain as if to commingle the visible and intelligible harmoniously. Did Titian co-opt at least part of his idea from the Greeks? Nearly two thousand years before him, Praxiteles was said to have sculpted a pair of identical Venuses—one fully clad, the other a nude. But the citizens of his island, Kos, offended by the undressed Venus, had the statue banished.

An earlier Renaissance work, Botticelli's *Primavera* (Figure 5-3b), painted about 1478, is a bit more complicated to interpret. It shows Venus at the center, mediating between the two loves. The carnal-terrestrial aspect—depicted in the form of the three figures Juvenescence, Splendor, and Abundant Pleasure—is positioned to the right, while love of the more ethereal kind is represented by the triad of Graces to the left—Chastity, Pulchritude, and Voluptuousness. Botticelli in effect has elaborated on the paired aspects that constituted the nature of love in Babylonian Ishtar's morning and evening star duality.

At the left side of the painting stands morning star Mercury. He turns from the real world with a passive kind of detachment as he brushes away the low-lying clouds—the mental nebulosities that only he, the guardian of the soul, can dispel. He only touches the clouds lightly, for they are delicate and transparent veils that can be parted only temporarily to reveal the true underlying wisdom about love. Once Mercury has removed the clouds of night (he agitates the wind, Botticelli tells us), at the other end of the canvas passionate Zephyr, the more tangible, earthly complement of Mercury, blows in from the west. He vigorously pursues—even lays hands upon—one of the innocent earth nymphs. As he touches her, flowers issue from her mouth, just as, with the arrival of the breezes of spring, cold earth bursts forth in bloom. Above Venus, in whose image the complementary halves of the Renaissance theory of love are united, lies Cupid, the mediator. He is the one who fires love's first arrow at one of the heavenly Graces to start the transformation.

Primavera is the drama of the rite of spring, the instant when the unattainable form of love (to borrow from Plato) is fully, yet momentarily realized in the florescence of new life on earth—and the planets, along with nature's other forces, are the star players. Botticelli has chosen to convey the celebration of the act of transformation in the form of dancing figures. "What descends to earth as the heat of passion returns to heaven in the spirit of contemplation," as art historian Edgar Wind puts it.[23]

This is the imaginative stuff of a rather complex intellect. The Venus love theme that Botticelli painted and that Ficino wrote about for their contemporaries was part of a great revival that took place in fifteenth- and sixteenth-century Italy. Painter and poet alike began peering into the pagan mysteries of the cosmos and trying to express what they imagined planetary images had looked like when their lofty Greek ancestors cast them in words and pictures, most of which had been lost in the Dark Ages. They saw themselves as reinterpreters and redevelopers of pre-Christian ideas about permutations and transformations.

In *The Birth of Venus*, also by Botticelli, painted after 1482, the celebration of the transformation process that brings about spring is once again the theme (Figure 5-3c). This painting—we might be tempted to call it *Venus on the Half Shell*—shows the Goddess of Love rising, like her Greek counterpart Aphrodite, out of the ocean. She lands in her shell boat on an imaginary shore, where she is greeted by Spring, who spreads out a welcome blanket of flowers, while the Zephyr's passionate breath spews out more flowers. But another mantle, one of protection, is extended by Chastity, and her posture is designed to display both her sensual and her virginal aspects. In another medium Petrarch describes this form of love:

> *And it is she*
> *we next observe, as, new born, she comes forth*
> *out of the sea, her lowly origin,*
> *as legend tells us. See how, lascivious,*
> *she rides within the cradling couch, adorned*
> *with crimson roses while swiftly flying doves*
> *provide her escort; mark her company,*
> *three naked girls, the first averts her face,*
> *the others outward look with smiling eyes,*
> *and all have snowy arms entwined in sweet*
> *reciprocal embraces.*[23]

Although these paintings come down to us as single images, many were intended to be viewed as cycles of permutations, the other parts of which have become separated or lost. Edgar Wind contends that *Primavera* may once have been part of an ennead, or ninefold grouping, that symbolized the conjunction of the sun (Apollo), Mercury, and Venus. Ficino, the teacher of Botticelli's patron, Lorenzo di Pierfrancesco, who commissioned the painting, tells us that the triad of planetary deities can be expanded into the ninefold powers represented by each member taken by itself and then in combination (in direct as well as reverse order) with each one of the other members of the triad.* This complicated notion reminds us of the way the astrologers of old believed they could discern the powers of the zodiacal signs by combining them in trines, squares, and so on. It is only the combination of Venus-Mercury that we happen to see in this painting; Venus dominates, but she is guided by Mercury; that is, the power of the soul lies at the center, aided and abetted by the power of the mind, and this reveals the ideal form of Venus-Aphrodite.

The transformation of the astrological pantheon into Christian imagery took place in literature as well as art. In the *Divine Comedy* Dante

---

* There is even a musical form for the ninefold way of joining the sun and his two closest companion planets. Each member of the ennead is a note in an octave, with the sun in the middle, in true Renaissance fashion. The ninth note is the unheard or highest note in the octave, which belongs beyond the planets, in the realm of the stars.

apportions the planets and the celestial spheres to a corresponding scientific discipline (Mercury is dialectic, Venus rhetoric). He creates a job description for each planet (Jupiter for rulers, Mars for soldiers), even an angelic choir (angels to the moon, archangels to Mercury). All innate human behaviors are hierarchically assigned to the arms of the blessed trinity. Dante's basic assumption—a reflection of the thinking of his day—was that nature was a harmonious ordered whole that was right with the world and with society. As in ancient times, the structure of astrological thought was predicated upon principles of hierarchy and alternation, in Dante's mind. All nature's aspects were interrelated as parts, with every entity—colors, plants, animals, protector gods, planets, day names, even parts of the human body—assigned its proper place. Like the ancient Roman mithraeum, Dante's hierarchy had a place for everything and everything in its place.

Unlike scientific discourse, in allegorical writing, poetry, and painting, the primary subject is kept out of view, disguised by secondary imagery. The reader or viewer is left to divine the intentions of writer or artist by uncovering hidden resemblances between primary and secondary subjects. In this sense, allegory is but another form of the unifying principle that characterizes the search for scientific truth.

The discovery process behind allegorical imagery and planetary mythological portraiture expressed in Botticelli's or Titian's paintings, Petrarch's poetry, Dante's writing, or Giotto's sculpture (Figure 3-1h) consists of matching concrete celestial images with human emotions and feelings. We might think it more mysterious and far less concrete than Newton's description of gravitation—perhaps because emphasis is given to the playful disguise that deliberately veils the intended underlying meaning. Its purpose, however, is to promote a human-centered discourse, not to give definitive scientific answers.

Today *humanity* implies freedom. The humanities we are taught in school focus upon shaping a life that passes beyond mere necessities, one that tries to overcome the dominance of culture by nature. Perhaps that is why the sciences and the humanities seem to conflict in their attitudes toward seeking truth. Such studies had their origins in fourteenth- and fifteenth-century Europe. They grew out of a tension between the beliefs that the course of human history was predetermined and that it could be markedly affected by the freely motivated actions of people. As you have seen, astrology, forever arguing that celestial movement directs our behavior, had already made peace with Christian dogma, which looked forward to the ultimate and inalterable Judgment Day. Both shared a fatalistic outlook. But the logic of Aristotle and Plato that emerged from a close reading of the newly recovered pagan documents had enthroned reason alongside blind faith as a way of acquiring knowledge and truth. The printing press was among many technological innovations that helped promote this inquiring attitude. For one thing, it helped make literacy more widespread and universal; for another, it facilitated a fresh exchange of ideas—it secularized knowledge.

The debate between the competing doctrines of faith and reason was widespread enough to become involved in politics. In Florence, for example, rationalism became the guiding philosophy of the Ghibellines, who sought to oust from power the more conservative Guelphs. They were suspicious of the inquiring spirit and clung steadfastly to church faith. Naturally, average citizens, for whom life was harsh, had little time to think about new ways of arriving at truth. Their real world was a series of hard-labor days from dawn to dusk, punctuated only by the occasional religious festival. But once Dante's and Petrarch's verse, written in the Italian vernacular rather than scholarly Latin, began to filter down to the public, the *rinascitá*, or rebirth as they called it, became a universal phenomenon. Wealthy banker Cosimo de' Medici spent his capital importing vast quantities of manuscripts from Alexandria and hired dozens of copyists and translators to make them available to teachers and students free of charge. The Medici family commissioned many of the works of art that explored and expressed humanistic themes. The effect of all this enlightenment was to veer public thinking in the direction of philosophy, nature, and contemporary society rather than religion, God, and the hereafter.

Sculpture, poetry, painting, and musical scores—not tablets, almanacs, and codices—became the texts of the early Renaissance humanists. Although in these media planetary imagery became a part of allegory, influence, affinity, force, and power—the concepts once discussed by Greek cosmologists and those that Renaissance scholars would grapple with anew—remained a part of the confrontation between people and the world around them.

The humanistic dialogue with the planets, which I have shown taking place in Renaissance art and literature, was not concerned with acquiring facts; rather it seems to have been about morality, human values—how we ought to behave as good Christians and good citizens. It served as a stimulus for new ways of expressing feelings about human desire and the nature and extent of our freedom. Can a person truly remain satisfied by tasting only the visible beauty of terrestrial Venus without succumbing to the lust that ensues when one indulges in the full range of sensual pleasures? Is it only by *contemplating* love that one can clearheadedly and intelligibly aspire to its truth? And how far can we exercise our freedom before we get into trouble? In a sense, we ask ourselves these same questions every time we seek the sensual pleasure of substances and stimuli that we know, if consumed excessively, can lead to our downfall.

THE SECOND DEATH
OF ASTROLOGY

Astrology died twice, really. As I have outlined, the beginnings of the first death of astrology in Western culture can be traced all the way back to St. Augustine in the early Christian era. He labored to point out the inconsistencies between Christian doctrine and the tenets of astrological determinism. The decline of Greek and Latin learning and the liberal arts, of which astrologers were a part, left few skilled, schooled astrologers in the world during the Middle Ages. "There is no pop without the philharmonic," says historian of astrology Jim Tester.[24] In other words, because there was no professional enclave of astrologers or official textbooks and tables, the art so popular in the pagan world fell into decline.

But astrology enjoyed a reprieve, thanks to its preservation and development in Islam, which blanketed many of the southern and eastern lands in the Classical Greco-Roman world, particularly places such as Alexandria, where so many of the old scientific documents were housed. Damascus and Baghdad became the new centers of higher learning, where Islamic scholars modified and improved the astronomical tables so necessary to the operation of a sound astrological system.

What enabled the cross-fertilization of Islamic and Western astrology? Love of mathematics penetrates to the core of Islam. God is unity, the highest number, and humanity ascends a ladder of multiple existences to get to him. Little wonder that once Islam encountered the Middle Eastern cradle of civilization after the fall of Rome, the Pythagorean and Ptolemaic manuscripts spoke to the conqueror with ringing clarity, for they were cast in the familiar and sacred language of geometry and number. Islamic scholars read and translated Euclid and Archimedes. They improved on planetary theory by relating geometry to algebra; they formulated the trigonometric functions, gave us their Arabic numerals and new instruments with which to chart the sky. They compiled twice as many celestial observations as the Greeks.

The same affinity that enabled the Moslem absorption of Greek science applied to Greek astrology. Because Islamic beliefs were attuned to allegory and symbol, Moslems were just as interested in genethliacal astrology, the casting of an individual's nativity horoscope, as their predecessors. Through astrology, they believed, one could discover his or her true cosmic dimension and know how this most lofty aspect of existence influenced daily life. Here was another way to achieve unity and order in the universe. Islam preserved and developed astrology because it was attuned to it. For a thousand years astrology lay well nurtured under astronomy's quantitative cloak in Cairo and

Baghdad, in Aleppo and Mosul, and when it was finally handed back to West Europe, in excellent health, the mathematical and observational accuracy of Islam had only enhanced it.

The same society that produced allegorical planetary imagery gave rise to another kind of confrontation with the celestial wanderers—one that will require an entire chapter to explore, for it is what ultimately killed astrology, or at least left it in its present lowly, unresurrectable state. Renaissance expressions of what the natural world was about echo from a tense time, when intellectuals who wanted to think and act more freely began to feel constrained by the demands of a deterministic universe, one that was animated, anthropomorphized—an environment in which matter and mind, body and spirit were one. They inherited a connected world that placed a premium on knowledge acquired from the macrocosm above because it was believed to have a direct bearing on what took place in the microcosm below—where these people lived and worked, loved and worshiped.

The freethinking humanists who began to shake the faith were partly responsible for astrology's second death, for under the same roof, mathematically based astronomical theory and human practice began to seem ever more irreconcilable. The bond between nature and culture was about to be severed by a new kind of common sense, although it would not happen overnight. Joining the humanists in the new spirit of inquiry were the empiricists, such as Bacon and Galileo, who began to view nature close up, to dissect it. They had begun to ask different questions: If the planets can act on our humors, how can they act on our bodies? Can we measure this activity? How can influence affect both individuals and societies? If the principle is not one of heat or light, then what is it? How do we make it tangible? Is there a mechanism? Over what time period does the influence function? Does it turn on and off or does it vary gradually?

And more questions: If the demon in the stars chooses my fate, is it not also possible that I might be free on occasion to choose the demon star? Or even if I can fully and confidently predict the course of Mars, how can I account for the subtle disparities among individuals ruled by him? How can two people born under the same stars turn out so different? Whereas celestial truths had once been accepted as revealed, radical thinkers of the age now felt the need to question assumptions and speculate upon them anew.

What Petrarch boldly wrote at the time of the Black Death in Italy hardly seems consistent with the popular belief that the great plague was caused by a planetary conjunction:

> *Leave free the paths of truth and of life. . . . These globes of fire cannot be guides for us. . . . The virtuous souls, stretching forward to their sublime destiny, shine with a more beautiful inner light. Illuminated by these rays,*

we have not need of these swindling astrologers and lying prophets who empty the coffers of their credulous followers of gold, who deafen their ears with nonsense, corrupt judgment with their errors, and disturb our present life and make people sad with false fears of the future.[25]

On the celestial battlefield Galileo's telescope and Copernicus's theorizing raised totally new ways of thinking about the universe. In the next chapter I am going to show how old ways of thinking were imperialized by the new scientific experimenters. Knocking the earth off its immovable pivot would have revolutionary implications in the continuing and ever-changing human dialogue with the planets. The spheres would be reimaged—old symmetries shattered, staid patterns of association between celestial and terrestrial things violated. New ideas about the size of the universe and the movement of its parts—and empirical demonstrations that these ideas make sense when the universe is viewed from a radically different, exterior perspective—would have a devastating effect on astrology. The "true" position of a planet would be given not by where it stood in the signs and houses but instead by where it lay in its orbit about the sun. Its "true" period of motion would become an entity we cannot witness directly from our skewed vantage point.

Empirical attempts were made to standardize astrology—to make it "sane," as reformers put it. This meant isolating natural from personal astrology, testing out weather predictions against conjunctions, seasonal variations against planetary appearances and disappearances, a patient's fever against the phases of the moon. But the very act of testing and evaluating, then re-forming predictions began to change astrology into a different discipline. The dialogue turned into a war of the worlds, and Petrarch's damnation of astrologers typifies the struggle on the humanistic battlefield of free will versus determination. To what degree do we have the power to set our own course of action? Can the shifting stars be entirely responsible for such devastation as we have witnessed?

On the contrary, the one who possesses scientific knowledge can walk on water, make the rains come and go, create ships that fly in the sky, and destroy enemy cities from great distances, or so Leonardo da Vinci proclaimed. His statements, which today seem startlingly prophetic, announced the betrothal of logic and reason to technology and machinery—a tenacious bond that still bears fruit, some of it admittedly too sour for consumption.

New ideas and new technology. The seeds of the confrontation between culture and nature in the form we experience it today were planted in late medieval and Renaissance times. By the time of Kepler (early seventeenth centruy), modern science's condemnation of astrology had begun firmly to take root—this despite the fact that Kepler was himself a mystic. For example, the seventeenth-century astrologer William Lilly is said to have predicted the Great London Fire of 1666; he was indicted for fortune-telling. He had issued a pamphlet twenty years before in which he noted that Mars, patron signifier of England, would move into Virgo, the ascendant to the English monarchy. This would happen at aphelion, the greatest distance of Mars from the sun. The outcome would be ominous to London, he said, to the merchants, to the sailors, to the common people, and it would come in dry brush fires and a plague. He was incorrect by a year in his initial prediction of the conflagration and the pestilence that followed. Only his friendship with an adviser close to Charles II got him off the hook.

Why this sudden change of attitude? Some of the history we read today characterizes the dawn of science as having occurred after the passage of the long, dark night of astrology, a blind alleyway paved with a seductive doctrine, absurd mythology, and inconsistently applied methods—an age when humankind was held captive by superstition and magic. How can modern people, who are well aware of how far away the sun, moon, and planets really are, find more attraction in thinking of Venus as a celestial marriage adviser than a fiery desert inferno blanketed by clouds, one debunker wonders. How can they follow such a monster, with scientific head and religious tail? He concludes: "We must speak out . . . to discuss the shortcomings of astrology and to encourage an interest in the real cosmos of remote worlds and suns that are mercifully unconcerned with the lives and desires of the creatures on planet Earth. Let's not allow another generation of young people to grow up tied to an ancient fantasy, left over from a time when we huddled by the firelight, afraid of the night."[26]

By contrast, anthropologist Stanley Tambiah points out the danger of reifying social behavior, that is, of categorizing enterprises such as astrology, alchemy, and magic into fixed, bounded systems and thinking about them outside of history. He argues that in its time the astrology of the early Renaissance was an imaginative and creative enterprise that fired the imagination of every educated person. It was a way of solving puzzles—a means of seeking answers to big questions. It is unfair of us to pass judgment upon it, to assign it a value out of its true historical context. Instead, we should be more concerned about astrology's way of explaining things for our predecessors who employed the system. Those who point out that many of astrology's predictions failed, or that they were cast on such vague terms as to leave a way out if they did not come to pass, neglect the harmonies, sympathies, and general worldview once attached to the system. They leave out the reciprocal and participatory orientation of the believer and focus only on the bottom-line test of the system, whether it "works" according to our definition of what it means "to work."

Even I will admit to having a personal astrologer. She is a charming elderly woman, a retired schoolteacher who lives here in town. Having attended my astronomy classes and open house lectures at the college observatory, she one day asked me whether I'd object to supplying her with the exact time and place of my birth so she could work out my daily forecasts. Ever the seeker and inquirer, I responded, and since then she has regularly produced for me packets of computer printout that cover several months of prognostication at a time. Today's entry says:

Transiting Neptune Sextiles Natal Sun
• *Meditate, dream, hope, be sensitive & reach out to the world around you.*
Transiting Uranus Sextiles Natal Mercury
• *Seek new contacts & friends who will help reflect your desire for new experiences.* [I expect shortly to have lunch with a new colleague.]

On the other hand, a few lines further down I read:

Act with discrimination. Your footing is not quite firm.

It is not favorable for you to state your intentions & gain understanding.

But finally,

Transiting Mercury Squares Natal Mercury
• *Make sure that you explain things carefully today.*

Transiting Sun Trines Natal Venus
• *You will feel at peace with the world. It's a good time to enjoy with others.*

Most of us find these indicative yet highly flexible statements entertaining; some think them all too facile to interpret, while others will go out of the way to avoid portended pitfalls. Still others shun such indicators. (Another astronomer here at the college has flatly refused the town astrologer's offer to cast his horoscope.)

Surveys reveal that most people who say they believe in astrology today actually interact with horoscopic charts at a superficial level, only browsing the astrology column in the daily newspaper. They know little of the complex, embedded alliance between the natural world of appearances and the ancient human imagination from which astrological forecasting has spun off. They are titillated by the element of mystery and attracted by the ambiguous nature of the messages, the possibility of having some measure of control in a complicated and chaotic world.

Like psychoanalysis, astrology offers a personalized, conversational framework. Except for a very small minority who regard witchcraft, satanism, astrology, and other occult beliefs as highly complex, deep structures that give real meaning to their lives, most people view with playful contempt what many once took quite seriously.

"Born in Cancer?" a tabloid column in a recent issue of the magazine *Vanity Fair* asks. Then you need not be concerned by friends and family who heap butter on their bread yet warn you about the dangers of a high-cholesterol diet when you do the same. (Does this mean because your sign is Cancer you are less subject to threat from saturated fats or less prone to damage your mental health by worrying?) Because your second house lies in a favorable position this month, don't worry about quelling your appetite. Librans should opt for getting together with their close-knit group because of a benevolent planetary traffic jam in the House of Friends.

The modern revival of the art of tabloid and commercial astrology began just before the turn of the twentieth century with magazines such as Britain's *Modern Astrologer* and Germany's *Zodiacus*, to which thousands subscribed, and the formation of cultic clubs such as the Astrological Lodge, founded in 1917 in London. In France a school of astrologers began to explore the statistical bases of celestial prediction. One of them, who called himself Fomalhaut after the bright star in the constellation of the Southern Fish, claims to have predicted the discovery of the planet Pluto in 1930. But Americans were already way ahead of the pack. By 1840 we had crafted our first horoscopic magazine. The first tabloid astrologers did not appear until the 1930s, and they seem to have had a rather odd inception. An astrologer was hired by the London *Daily Express* to do a horoscope for newborn Princess Margaret, daughter of the future George VI. As it turned out, he did a fair job of predicting her marital problems. He tossed in a few predictions for those with birthdays in the same week. Public reaction was so enthusiastic that it was not long before the paper developed a daily column on personal astrology (the *New York Post* followed a few years later). Today's mass market includes hundreds of journals, nearly two thousand tabloid columns, and dozens of "Dear Abby"–type horoscopes in fashion and travel publications, not to mention several pay-per-call phone numbers and personal computer software.

Even a nonbeliever cannot fail to be drawn in by the playful quality of these sorts of modern omens.* Today's astrology offers something ambiguous and open to interpretation in an age that purports at every turn to give definitive answers to any question we ask.

How, then, do we account for the fervor of many a modern scientist turned astrological derogator? Why has astrology remained a monkey on the back of science? I am reminded of the passion with which the

---

* A popular theory in Japan places blood type above astrological sign. Type A blood, for example, predicts you will be cautious, eager to please, indecisive, sympathetic, and conformist. This is an interesting example of mainstream divining that coopts scientific principles. Most Japanese scientists, needless to say, are nonbelievers.

scientific orthodoxy overreacted to Immanuel Velikovsky's popular theories four decades ago. In *Worlds in Collision*, he postulates a series of bizarre cataclysmic scenarios involving the planets based on a literal reading of the Bible—the earth stopped rotating; Venus, torn from the bowels of Jupiter in historical time, flew like a comet close to the earth, and so on. Scientists' efforts to ban his books only seemed to accredit Velikovsky's claims against the establishment—at least in the eyes of the public. For the scientists these exercises in renunciation seem, at least on the surface, to constitute a reaffirmation, a way of celebrating the correctness of their way of understanding nature.

Perhaps science education is at fault. Not content to recognize that astrology simply does not operate under the same logical principles and presuppositions we use when we deal with the unseen world by the scientific method, we lash out against the handful who still practice the occult art and condemn them for having failed to progress beyond an outmoded way of thought. Even laypeople who have little understanding of the intimate workings of scientific methodology relegate the astrologer to the same ignorant class as our so-called dim-witted ancestors. Too weak to realize the power of their own will, they, like our ancestors, force themselves under the yoke of nature's whim.

You cannot know nature today simply by worshiping its extraordinary powers. Valuable knowledge, as we now come to regard it, has become a kind of truth conceived *in fact*, acquired through a process that encourages an incomplete, never adequately formed yet empirically testable worldview. So different from astrological tenets is the scientific process! It remains ever flexible, forever leaving matters open to question and theories available to further perfection in its gradual ascent toward a universal order that we already have faith can never be fully realized. Now this regimen of living within a progressive, forever changing worldview is filled with psychic stress. Believing in science carries with it a very heavy burden of uncertainty, for we must always allow for the possibility that we really might be influenced by indefinable forces beyond our control. In our weakest moments, we collectively find ourselves even more helpless to act than the old fortune-tellers, who at least could identify the demons and ghouls they were dealing with.

The medical establishment has criticized medical gurus who, lacking faith in Western medicine, appeal to potential elements within us to overcome by our own free will a prognosis destined to befall us. The process is similar—the denial of belief in the system in vogue (science) and its replacement with one that is now clearly out of fashion (astrology). Do the debunkers overexert themselves in the defense of science, not because they feel enthusiasm about celebrating its successes but because they lack the quiet confidence that it can really solve all our problems?

I have explained, then, how the system of astrology in use today has its logical underpinnings in ancient theories of planetary motion and obsolete ideas about the interrelationship between matter and spirit. Whether it meets our modern criteria for common sense is unimportant, for it never was intended to apply to us. What *is* important is that astrology always was a rational enterprise; it had structure, interconnected parts, and a set of rules that showed how the parts were connected. The structure of astrological thought was based upon principles of hierarchy and alternation, of time-free association, sign, and meaning, not chronological cause and effect.

Astrology used empiricism to reinforce its categories. Many of the habits and characteristics of mythological participants in the drama of life were based on very real observations designed to appeal to the visual sense of the believer, but beneath the abstract symbolism lay the penetrating gaze of Mercury the prophet and the resurrected Venus, who emerges from the underworld. Sometimes these mythologies became detached from the empirical world of hard facts, to which we tend to give primary attention today, and developed independently. But, for those who seek it, the real world of the senses always remains embedded within.

Although I did not set out to write a history of astrology, I trust I have shown that believing in the stars, in Western culture, has a deeply rooted, ever-changing history. That it survived so long is testimony to its success as a medium for dealing with the great human questions. Stellar determinism was also a vital part of other past civilized cultures of the world. Although we have no detailed accounts of the history of non-Western astrologies, they too seem inextricably tied to astronomies. People first follow the stars, then they come to believe in them. The only violation of this universal law seems to have occurred in a radical turn taken by Western culture several generations ago.

CHAPTER 6

TECHNOLOGY:
HARNESSING THE IMAGES

This prying tube too shews fair Venus' form
Clad in the vestments of her borrowed light,
While the unworthy fraud her crescent horn
Betrays. Though bosomed in the solar beams
And by their blaze o'erpowered, it brings to view
Hermes and Venus from concealed retreats;
With daring gaze it penetrates the veil
Which shrouds the mighty ruler of the skies,
And searches all his secret laws. O! power
Alone that rivalest Promethean deeds!
Lo, the sure guide to truth's ingenious sons!
Where'er the zeal of youth shall scan the heavens,
O may they cherish thee above the blind
Conceits of men, and the wild sea of error
Learning the marvels of this mighty Tube!
—JEREMIAH HORROCKS, QUOTED IN ARUNDELL
WHATTON, P. 50–51.

UNVEILING VENUS

When Copernicus challenged the ancient Greek tradition of geocentrism by daring to propose that the earth revolved about the sun, he did so because, as he said, "I think it is *easier* to believe this than to confuse the issue by assuming a vast number of spheres, which those who keep the Earth at the enter of the universe must do. I thus rather follow Nature, who producing nothing vain or superfluous often prefers to endow one cause with many effects."[1] Like all great scientific revolutions, Copernicus's ideas were born out of the tension that always exists between the forces of

178

tradition and those of innovation: Darwin countering the biblical view of creation, Freud arguing that the power of the unconscious was a greater source of human motives than the conscious, Einstein positing relative frames of reference to replace the absolute.

Imagine how foreign Copernicus's declaration would have sounded to a practicing astrologer in midsixteenth-century Poland. Copernicus, like the astrologer, still followed nature—except he did not do what the stars told him. Instead he looked beyond nature's imagery and into its underlying way of behavior. His questions dealt neither with divine spheres and circles nor with signs and meanings but with operative causes and their effects. Gone were the vital spirits of the flying wanderers; they were now steered by forces outside themselves. In the future the planets would take up their courses without a care for humanity.

Copernicus was born in Toruń, a turbulent frontier town on the Polish border with Prussia. There Knights Templars, a military and monastic order dedicated to maintaining free passage for pilgrims to the Holy Land, had returned home disenfranchised after the unsuccessful Crusades to wage a territorial war of conversion against the local heathen population. Having enrolled at the University of Kraców to study religion in 1492, the year Columbus came to America, Copernicus acquired the new spirit of humanism, sharing in the rediscovery of the old Classical texts and plumbing their depths. University life then was nothing like it is today. We can imagine Copernicus conversing in Latin with his fellow students and debating with his professors (a college requirement) after a hard day that began at 7:00 A.M., listening to their lectures on Euclid, astrology, geometry, and philosophy—all taught from a medieval point of view and all heavily directed toward preparing men for religious work. The career path Copernicus took, which carried him to the great Italian universities, also included canon law and medicine—a good churchman needed to know about principles of healing if only to discern quackery from legitimate practice, lest how could he administer to the total spiritual life of his subjects?

Armed with one of the best educations in Europe, Copernicus returned to his native country and became canon of Frauenburg (Frombork), which meant conducting services in the cathedral and managing estates that belonged to his chapter. There he reformed coinage, wrote tax laws, even developed sophisticated physician's skills. Had Copernicus not been so underworked in his position, he might never have written the great book that knocked the earth off its pivot.

Truth is simplicity. Although the world may look very complicated to us, it is really quite simple, provided we understand it in the proper way. This article of faith was the credo behind the questioning attitude inculcated in Renaissance schools. Applied to the problem that interested Copernicus, it reads: Could the movements of the planets, which seemed so complex according to the earth-centered model of the solar system, be explained in a simpler way by supposing that the sun instead lay at the center?

The Greco-Babylonian way of understanding the universe, freshly resurrected in the Renaissance, laid heavy emphasis on the value of the mathematical and geometrical prediction of events that happen in the real world. The mother tongue of the new astronomy was the quantitative description of the movement of celestial bodies in spatial coordinates. Its language, true to tradition, consisted of planes, orbits, and nodes (intersections). The principal issue concerned whether to place the axis depicting the center of coordinates through the earth or the sun. Which centrality would reduce the depiction of the observed motion to its most economical and elegant form? The answer could be arrived at only with an enormous amount of mathematical effort. Imagine the difficulty of calculating the sizes and shapes of the planetary orbits from actual observations—all by hand with pencil and paper!

By discovering a formal mathematical structure that underlay the movements of all the planets, including the earth, Copernicus achieved satisfaction by creating order out of the seemingly chaotic planetary motions he and his predecessors saw in the sky. True, some planets' orbits are now known to be slightly more or less elliptical and others might have higher or lower tilts relative to the plane of the earth's orbit about the sun. Nevertheless, the overarching perspective that unites the planets in Western astronomy is that although they *appear* to move in a complicated way among the constellations, in fact they *really* move about the sun in a much simpler way. We have come to be satisfied with such an interpretation even though it requires an immense mental leap that carries our minds off the surface of our planet.

Not to lose perspective in the great story of planetary imagery, I should pause to point out that most people who have inhabited our planet have not really cared about this problem. And the languages others spoke in their dealings with nature were somewhat different. Some societies were not even impressed by "planet" as a category among celestial phenomena the way we define it or by the likenesses that our Mediterranean predecessors sought between the motions of Mars and Venus or Mercury and Jupiter. Any desire they had to describe how planets move in space seems disconnected from any question about what physical body lies at the center of motion.

If Copernicus's wild idea was true, its repercussions would quake the foundations of religious, philosophical, and physical belief. Would God create a garden for humanity to tend in so remote and ordinary a place as this? If the world turned instead of the sky, should we not all fly off into space? Would birds and clouds floating in the air not get left behind? And if the whole sky did not rotate in a day, who knew how far the other worlds we see in it could really be? Could our vision penetrate to infinity rather than to the well-guarded limits of a protective canopy in the sky? What could those other worlds out there be like? Were they stars like the sun? Inhabited worlds like our own?

It was not Copernicus's idea alone that catapulted the sun-centered universe into the mainstream of our modern belief system. Empiricism—technologically aided observation—played as vital a role. The inquiring spirit that asked all those questions also queried: How can we see this model work? This chapter will demonstrate how a single instrument in the hands of two contrasting empiricists changed the images of the planets forever.

Galileo was a central figure in planetary astronomy who was swept up in the excitement of Copernicus's new ideas. He lived and worked in the thick of the great intellectual battle that ultimately overturned the way we think of the world and our place in it. Einstein praises Galileo as much as we now praise Einstein.[2] He calls him a courageous individual who stood up against those who relied on their authority and position in the intellectual hierarchy to promote their version of the truth. We revere Galileo and his contemporaries because we glimpse in Galileo's endeavors the thinly veiled structure of our modern way of knowing the natural world. Reality legitimates belief, and he delivered us a fresh view of the real world close up through the looking glass.

Galileo's observations of nature became the key testimony in the great planetary revolution. They appeared in a tidy little book, *The Sidereal Messenger*, which he published in 1610. An observational diary, it contains clues that first opened our modern minds to the possibility that we inhabit a limitless universe, for by removing the earth from fixed centrality, as Copernicus had done, Galileo was able to apply his observational evidence to validate other ways of explaining celestial motion without constraining everything in the sky to a necessarily small sphere that turned about us in twenty-four hours. We are still trying to cope with the religious, philosophical, and moral effects of living in such a place. If we are detached from the planets, why were we put into such a vast universe? If we are not here to serve nature and to have nature serve us, then what is our purpose? What is our mandate?

Galileo's writings teach us to question, to trust our senses, and, above all, never to accept blindly what our betters tell us simply because of their position in a hierarchy. For the seventeenth century, this was radical stuff, and it got him into deep trouble with the church fathers.

Most people who have heard of him think of Galileo as the founder of modern astronomy, the first person to behold the sky through a telescope, a brilliant and indefatigable role model for the modern scientist, who, with that lean-and-hungry look, quests selflessly in the lab today for the betterment of humankind tomorrow. Indeed Galileo was a man whose tenacity and conviction drove him to present to the church leaders ideas he knew they would find morally and socially unacceptable—a public act that had grave personal consequences.

Born in Pisa in 1564, the same year as Shakespeare, Galileo had built a fairly undistinguished career as professor of mathematics in Padua. Above all he was a very careful and astute experimenter, having conducted studies on falling bodies, the tides, and the pendulum. He was interested in just about everything in the physical world—the tides in Venice, military engineering, magnetism, falling bodies—he even invented an early form of the thermometer. To be sure, Galileo *was* a first-rate scientist. However, time has a way of air-brushing away a hero's flaws. As science historian Albert van Helden tells us in his very humane introduction to *The Sidereal Messenger*, Galileo was every bit as concerned about his own betterment as any hardened professional of our own age—about getting to the scene of discovery before anybody else and launching his findings into print before somebody "scooped" him. He was rather ambitious, forthright, even polemical.

In 1608, already well into his forties, Galileo learned that a lens maker in Holland had built an "optic reed"—"a certain device by means of which all things at a very great distance can be seen as if they were nearby."[3] Actually, Galileo was one among a host of individuals who inquired about the gadget, but what set him apart from his colleagues was that he was clever enough to use diplomatic channels to acquire the recipe for constructing one. He obtained a pair of spectacle lenses (the term comes from *lentil*, whose shape each piece of glass approximated), placed them at opposite ends of a hollow tube at the sum of their focal lengths, and thus fashioned his own spyglass.

Soon Galileo learned to grind and polish lenses on his own so that he could improve the instrument by experimenting with different components. His colleagues, those tradition-bound professors Einstein berates who often seem afraid of what new ideas might do to their job descriptions, were naturally reticent to look through his sight tube. They thought that it would distort the "real" world. We cannot blame them, for their common sense held that all objects emanated "species" or perfect likenesses of themselves regardless of whether an observer was present. And Lucretius had taught fifteen hundred years before that optical images are created by thin outer skins that perpetually peel off the surfaces of bodies, fly through the air, and enter our eyes directly. Imagine how the glass lens would distort the reality conveyed by these delicate flying films once they splattered upon the surface of the receiving end of an optic reed. "We do not admit that the eyes are in any way deluded," said Lucretius.[4]

Even before design flaws were revealed in the Hubble Space Telescope, some had pronounced the billion-dollar instrument the greatest potential boon to astronomers since (and possibly even before) the telescope of Galileo, relatively speaking. This may prove true, though I find it doubtful, for unlike the Hubble, the Galilean telescope came along at precisely the right moment in history—when innovative ideas, fully formed, were ripe for empirical testing. Western science was ready for a novel strategy: to penetrate nature, to take it to pieces in order to get at the subtler, finer details beneath the surface, the irreducible facts where real Truth—with a capital T—lay.

Before the Renaissance, truth was arrived at by contemplation rather than by manipulation, a more passive form of acceptance rather than intervention in nature's processes. Within a generation of Galileo, Sir Francis Bacon would produce the *Great Instauration*, a manifesto that called for human action in nothing less than an all-out assault on nature. His agenda was to probe and ultimately conquer nature. Baron, viscount, member of Parliament, lord chancellor—this Cambridge-educated man of the law saw fit to organize a method for acquiring knowledge of the physical world. Bacon's program reads almost like a set of traffic rules telling what each discipline should contain, what data should be acquired, what ought to be dismissed as foolish. The greatest promise of all, he argues, lay in technology, for only by extending the senses could we unmask nature's deeper and darker submerged secrets. The new philosophy, framed in a mechanistic way of thinking, would be driven by an aggressive optimism that everything from music to morals could be cast under the umbrella of natural law. The new science would be spread about the literate and educable public via demonstrations, popular lectures, magazines, and libraries. It was the beginning of the Age of Enlightenment and of the demise of planetary astrology.

Bacon's futuristic instrumentation catalog seems almost clairvoyant, prophesying everything from lasers to cryogenics, air-conditioning, and stereo sound. He tells us, for example, that the future would produce new artificial metals, forms of refrigeration for curing diseases so that we could live very long lives, then continue life by preserving parts of the body. We would conserve the snow; make fresh water from salt; build engines that multiplied the winds; cause garden vegetables, trees, and flowers to germinate earlier; develop new kinds of fruit. We would multiply light and make it sharp so that it would penetrate to a great distance, make small sounds into great and deep ones, subtle smells into more potent ones, and disguise certain unpalatable yet nutritious foods so as to make them more appealing to our taste—and we would make bigger and better cannons as well. All of this he wrote in the early 1600s. His ultimate goal: the mastery of nature for the good of humanity.[5]

As physicist Hans von Baeyer has pointed out, Bacon's aggressive attitude toward nature, exemplified by "putting her to the question," was very much in tune with the methods he employed as an Elizabethan lawyer for extracting information from his witnesses—inquisition by torture. Perhaps we can understand why Galileo's colleagues were suspicious when he jabbed the sky with his optic reed.

Fueled with the Baconian attitude and thanks to his own ambition and initiative, Galileo quickly pulled well ahead of the local technology. He went before the doge of Venice to demonstrate his product. He impressed the court and even donated his tube to the state, but not before making his point—the same sort of pragmatic case supporters of the space program made to the U.S. Congress back in the 1960s—that the military advantage such a piece of equipment would give the republic (of Venice) was incalculable. As van Helden candidly puts it: "In other words, Galileo gave the Doge and the Senate sole rights to the manufacture of his instrument and in return he asked very tactfully for some improvement of his position at the university."[6] Evidently Galileo, like all struggling academics, needed the cash. He had to support his sisters' dowries, not to mention the two daughters and a son he had fathered out of wedlock. He had already been renting out rooms to students and running a little business in the manufacture and sale of small scientific instruments to augment his meager salary.

Galileo was driven—driven to make a bigger and better "optic reed," and by the end of 1609 he had produced an instrument capable of revealing what no human eyes had ever seen in the heavens—mountains on the moon, dark spots on the face of the sun, and satellites around Jupiter that moved (see Figure 6-1a):

> *Accordingly, on the seventh day of January of the present year 1610, at the first hour of the night, when I inspected the celestial constellations through a spyglass, Jupiter presented himself. And since I had prepared for myself a superlative instrument, I saw that three little stars were positioned near him—small but very bright. Although I believed them to be among the number of fixed stars, they nevertheless intrigued me because they appeared to be arranged exactly along a straight line . . . But when, on the eighth, I returned to the same observation, guided by I know not what fate, I found a very different arrangement. For all three little stars were to the west of Jupiter and closer to each other than the previous night, and separated by equal intervals.*[7]

He also witnessed strange appendages next to Saturn (they turned out to be rings):

> *The star of Saturn is not a single one, but an arrangement of three that almost touch each other and never move or change with respect to each other, and they are placed on a line along the zodiac, the one in the middle being about three times larger than the other two on the sides.*[8]

And the planet Venus revealed itself as a disk that passed over many months through a cycle of phases just like the moon:

I began to observe Venus with the instrument and I saw her in a round shape and very small. Day by day she increased in size and maintained that round shape until finally, attaining a very great distance from the Sun, the roundness of her eastern part began to diminish, and in a few days she was reduced to a semicircle. She maintained this shape for many days, all the while, however, growing in size. At present she is becoming sickle-shaped, and as long as she is observed in the evening, her little horns will continue to become thinner, until she vanishes. But when she then reappears in the morning, she will appear with very thin horns, again turned away from the Sun, and will grow to a semicircle at her greatest digression.⁹

Crescent-shaped planets, planets with rings and moons. It seems uncanny that even the crudest telescope, such as the one Galileo used—a mere pair of shaped pieces of glass set apart by a long cylindrical tube—is capable of revealing the visible universe to be such a remarkably different place from the one that confronts the unaided eye. Jupiter's satellites, for example, are just at the limit of naked-eye visibility. Figure 6-1b, which shows what it looks like through a telescope, casts Venus, like the image of Jupiter in its companion illustration 6-1a, in a very unrealistic background. Some have even speculated that careful sky watchers who look at dark skies through unpolluted air can actually see the *crescent* Venus without optical aid.

Venus, in the incarnation of Ishtar-Astarte-Aphrodite, is often pictured riding a bull, and she has been related as well to the horned altars of Knossos on Mycenaean Crete. Recall that in Babylonian mythology Ishtar is the daughter of Sin, the crescent-shaped moon god. Did keen-eyed observers in the clear air of Mesopotamia actually see the resemblance between father and daughter 3500 years before Galileo detected it with the new telescope? Some quotations from the cuneiform literature are highly suggestive:

> *If on the right horn of Venus a star is visible you*
> *will have good crops in the land.*
> *When upon the right horn of Venus a star is not*
> *visible the land will bear many misfortunes.¹⁰*

> *If Ishtar takes away upon her*
> *right horn a star, and if Ishtar is large but the star*
> *small, the kind [people of the city] of Elam will be*
> *strong and mighty.¹¹*

The ancient Mexican codices may have described Venus as a stippled ball, sometimes accompanied by a pair of horns that give it a crescent shape (Figure 3-4f). It is difficult to attribute all these associations to

FIGURE 6-1. THE PLANETS AS NEVER BEFORE REVEALED
Donato Creti's Planetary Series (1711).
6-1a. The planet Jupiter is shown in the skyscape but as astronomers on the landscape below would have seen it with their telescopes —a round disk with cloud markings accompanied by a retinue of satellites (three of the four Galileo observed are shown).

6-1b. Crescent Venus, seen through a telescope, imitates the moon in the sky. Note the woman in the foreground.

These were part of a set of drawings intended to decorate the entryway of the Vatican Observatory. (Musei e Gallerie Pontificie, Vatican City)

lucky guesses, especially when we find that they offer reasonable expressions of astronomical symbolism in different cultures.*

Crescent controversy aside, Galileo pursued his observations of Venus with different motives from those of Ishtar's worshipers. Yet he still applied the feminine gender when discussing the planet (and the masculine one to Jupiter), for the separation between the physical and the metaphysical—between astronomy and astrology—that we take for granted today was but a tear in the fabric of the cosmic worldview four hundred years ago. Little wonder, then, that the man who wrote the statement I quoted about his attentive observations of Venus also composed this rather animated passage on life, light, and spirit in the sun:

> *The Sun is a meeting point in the centre of the world for the lights of the stars which are spherically placed around it. They emit their rays which meet and intersect in the centre where they grow and increase their light a thousandfold; so that the light thus strengthened is reflected and spreads itself rather more vigorously, full of virile, so to speak, and lively heat, and so gives life to all the bodies which orbit its centre; so that it is rather like the heart of an animal in which there is a continual regeneration of the vital spirits, which sustain and give life to all its members . . . just as the Sun, while nourished from without, sustains the source from which this light and prolific heat continually emanate, which gives life to all the bodies which surround it.[12]*

This is no simple metaphor. For Galileo, life and matter were still very firmly connected.

Of his Venus observations, Galileo wrote in a letter to a colleague: *"Haec immatura a me iam frustra leguntur o y."* Of course, it was the custom to communicate in Latin, the universal scholarly language, but here Galileo seemed to be writing nonsense, for it says: "These unripe matters are brought together by me in vain." Galileo had chosen to communicate with anagrams to safeguard his own discoveries lest someone else claim them. The Italian scientist even refused to loan one of his telescopes to Kepler, his German ally who was also active in promoting Copernicus's ideas.

Kepler tried to decipher the Venus anagram by unscrambling all the letters. What he came up with is impressive though not correct: *"Macula rufa in Jove est gyratur mathem,"* which translates roughly as "There

---

* You may wonder: If the Maya or the Babylonians also saw a crescent Venus, is there any evidence to suggest they believed the sun shining on different parts of the planet caused its phases? The answer is probably no. Because these cultures did not think of the universe in terms of physical bodies orbiting one another in empty space, it would not have been in their nature to make the same deductions Galileo did about whether the earth or the sun lay at the universe's center. After all, remember that even the Greeks thought the planets were self-luminous spheres immersed in a crystalline medium.

is a red spot in Jupiter which rotates mathematically." In his return letter to his Italian colleague, he pleads: "Please don't withhold the solution from me; you're dealing with an honest German."[13] Galileo's intended message was *"Cynthiae figuras aemulatur mater amorum,"* or "The mother of love emulates the figures of Cynthia"; in simpler terms it means that Venus imitates the phases of the moon (Cynthia). Our twentieth-century science books are already unintelligible to so many people—imagine compounding the difficulty by communicating theories via anagrams written in a foreign language!

There were even deeper reasons for Galileo to veil his observations in secrecy, for, as did his other discoveries with the telescope, the visible revelation of the phases of Venus carried rather controversial implications, especially for the church, which, for reasons already obvious, opposed the ideology that the sun lay at the center of the universe.

Just why is this Venusian evidence so seminal in removing the earth from its fixed pivot? According to Ptolemaic theory, which had carried the day for more than a thousand years, Venus traveled on an orbit called an epicycle, the center of which revolved about the earth and also lay on a line between earth and sun. Assuming Venus receives its light from the sun, there is no way a terrestrial observer can account for the observed elongation or side-to-side motion of Venus relative to the sun and the full cycle of phases at the same time if the earth is placed at the center of motion. But what Galileo saw through the telescope with his own eyes *can* be explained by holding the sun fixed and putting the earth and Venus on different orbits about it, fast-moving Venus on the inside track and the slower earth on the outer orbit. This is the only sensible way Venus can expose us to all the phases from crescent to gibbous to full then back again to crescent, as Galileo had observed.

These simple facts of observation fit in an elegant way with Galileo's other telescopic revelations: that the moon, with its mountains and pitted surface, looks like another world; that Jupiter imitates our sun by serving as the center of its own family, consisting of several smaller bodies that orbit about it; that the hitherto unblemished solar deity could be thought of as simply another celestial sphere because now Galileo had seen spots upon its face. All these fresh facts uncovered with Galileo's spyglass conspired to weaken any belief in detaining the earth at the center of a universe filled with pristine fixed points of light. As Galileo sardonically put it to those stodgy professors he loved to taunt: What a powerful instrument the telescope is. The uncompromising evidence it brings to our eyes immediately explodes all the disputes that have tormented philosophers for ages.

Not only did Galileo bring technology to sky watching but he also altered its agenda. Now one could ask questions about the physical makeup of the celestial bodies, such as Jupiter and Mars, the sun and moon, instead of just following their motions and describing them mathematically. And from now on anyone who picked up a telescope and looked skyward would be compelled to think he or she was probing a vast, perhaps limitless volume of space in which our world, at least materially speaking, had become an insignificant speck. The Venus Galileo's predecessors had attentively followed across the heavens was no longer just a dazzling white light. Those who faced it close up under glass saw a countenance that revealed changing expressions. But Galileo paid the price of both his freedom and his honor for the way he chose to interpret his telescopic observations. He was tried by the church for preaching the false ideology that the sun and not our world of Genesis lies at the center of the solar system. Even after being forced on bended knee to admit that he only preached the Copernican theory as hypothesis, he was confined to house arrest.

In a long letter to Christina, grand duchess of Tuscany, Galileo refutes the charges, touching on scripture in much the way we might expect of one of his modern scientific counterparts: "Would they [the holy fathers] have us altogether abandon reason and the evidence of our senses in favor of some biblical passage . . . ? [Is not] the intention of the Holy Ghost . . . to teach us how one goes to heaven, not how heaven goes?"[14] In the letter he tries to separate questions of ethics from those of science, rational explanation from the revealed world. Both schisms would ultimately tear at the fabric of astrology, which sought to connect humanity firmly to the physical world.

ODE TO THE TUBE

The epigraph at the head of this chapter was composed, scarcely a generation after Galileo's penetrating gaze, by a little-known twenty-year-old genius, an amateur astronomer from England named Jeremiah Horrox (or Horrocks, as it was later spelled). He wrote those lines shortly after he, too, had viewed Venus through a telescope. The words are worth contemplating for two reasons: First, they were written at a time when the Western world was just beginning to appreciate the power of technology to enhance our senses. Horrocks's words demonstrate an awe of and passion about the then new high-tech instrumentation that we tend to take for granted. Second, his words—more than Galileo's—reveal the tension between reason and revelation that still veiled the planet watching of that time.

The occasion was the first recorded observation of a transit of Venus across the surface of the sun, an event that occurs on the average only once or twice a century. (A pair of such events will next happen in 2004 and 2012; the last one took place in 1882.) Mercury, Hermes in Horrocks's ode, had already been observed in transit a few years earlier, its "concealed retreats . . . penetrate[d]," as Horrocks says in poetic language. But young Horrocks was the first to predict mathematically the far-rarer occurrence of the passage of Venus in front of the sun.

He had just been ordained curate of Hoole, near Preston (north of Manchester-Liverpool), another astounding achievement for one so young, but he devoted every spare moment to astronomical pursuits. The inspired new clergyman gave this as his motive for practicing astronomy: "It seemed to me that nothing could be more noble than to contemplate the manifold wisdom of my Creator, as displayed amidst such glorious works; nothing more delightful than to view them no longer with the gaze of vulgar admiration, but with a desire to know their causes, and to feed upon their beauty by a more careful examination of their mechanism."[15] Horrocks's philosophy is based on the premise that the ascent of science begins with the ascent of humanity, whom God made for the express purpose of explaining and interpreting the world. Our mandate is to understand in depth all the treasures God created, for we and we alone are served by all creatures and are the servants of none. With animated conviction about the bond between God and nature, Horrocks used every pittance he could acquire to purchase astronomy books—he called them his teachers and his weapons—to set about discovering his Creator's secret laws.

Horrocks was innately so skillful at mathematics that it did not take him long to perceive the defects in his arsenal. One of the first to recognize and apply the significance of Kepler's writings to planetary computations, Horrocks soon found himself using these new data to update and correct the errors in standard published tables of position that professional astronomers all over Europe had been using for decades without question. And it was while he was predicting the future positions of planets that his computations revealed that within a month there would be a conjunction of Venus and the sun. According to his calculations, Venus was due to pass across the solar surface on the afternoon of November 24, 1639. Horrocks wrote to his friend William Crabtree, a draper who lived near Manchester and shared his pastime through regular correspondence. The two planned, in their respective locations and by the most effective method, to observe the event: It would be best, Horrocks thought, to project the image of Venus through the telescope onto a screen in a darkened room, the way it is pictured in Figure 6-2.

A Venus transit lasts about six hours, but, for much of the first half of it Horrocks's notebook shows no record of his having observed the event. In fact, his diary tells us he was called away to more important business. He does not say what it was, but a biographer tells us that the twenty-fourth of November fell on a Sunday in 1639, and the transit began at 10:00 A.M. He speculates that Horrocks was probably called

FIGURE 6-2. VENUS UNVEILED
William Crabtree observes the transit of Venus across the face of the sun by projecting the image onto a screen in a darkened room. (City of Manchester)

away to his devotions and duties at church and for this reason missed his observations.[16]

By Sunday afternoon, when Horrocks returned to his instrument, the transit was well in progress. A short bout with the clouds had passed when the young amateur astronomer witnessed his prediction come true: "I then beheld a most agreeable spectacle, the object of my sanguine wishes, a spot of unusual magnitude and of a perfectly circular shape, which had already fully entered upon the sun's disc on the left."[17] Attempting to keep a reasoned head, young Horrocks went on to describe the physical elements in detail: the angle and inclination of the path, the diameter of Venus, the estimated distance between the center of Venus and the center of the sun at different times as the transit progressed.

In triumphal tones, Horrocks expressed his elation. He had just emerged with his weapons—his eyes, his books—victorious from battle:

> *Hail [then] ye eyes that penetrate the inmost recesses of the heavens, and gazing upon the bosom of the sun with your sight-assisted tube, and dared to point out the spots on that eternal luminary! . . . Contemplate, I repeat, this most extraordinary phenomenon, never in our time to be seen again! the planet Venus drawn from her seclusion, modestly delineating on the sun, without disguise, her real magnitude, whilst her disc, at other times so lovely, is here obscured in melancholy gloom.[18]*

Triumph over the inconstant wanderer—Venus unmasked—the truth revealed. Strange words and endeavors for a churchman, even a young, eccentric one prone to burst into poetic strains. The youthful genius–biblical cleric seems engaged in a vocal dialogue with Venus as animated as that of a Babylonian priest addressing Ishtar. One major difference, however, is that Horrocks praises himself and the tools of his craft as much as the spirit who fabricated the natural laws that now structure the universe. His words elevate human beings at the expense of their deity.

If the dual interests of praising God and probing nature seem incompatible, remember that, up to this time, European culture had not yet breathed an atmosphere that separated science from religion. Today knowing God and knowing nature for most of us constitute two distinct agendas—one ruled by spirit, the other by matter; one of faith, the other of reason. But sixteenth- and seventeenth-century European Christians did not think and act as we do. They were radically shifting their perspective on the interrelationship between matter and spirit, from a worldview in which God intervened directly and served as both the immediate and the final cause of all natural processes and events, each with its own purpose for us, to an outlook in which God created the universe, set up the laws that would govern its operation, and then loosed his well-tuned machine to hum along in an orderly way. Praying to God does not induce rain. It rains when it has to rain because that is the way God set up the rain cycle in the first place. In a sense, the revolution in which Galileo and Horrocks participated was still as much about the nature of God as it was about the nature of nature.

The God young Horrocks addresses seems a more remote, less accessible deity who works in far more abstract and indirect ways than the transcendent Yahweh of Genesis or Tohil of the *Popol Vuh*, whose very spoken words were believed to be immediately translated into action. To understand the wisdom of the God of the Renaissance, one needed to investigate the mechanism he created, to probe the causal laws he set up that underlie the mechanism, to manipulate the complex mathematics by which he caused those laws to operate, and, as Bacon advises, to gaze far more penetratingly upon the manifold wonders that God's wisdom produced and that only our eyes were destined to see.

Incidentally, as Figure 6-2 depicts, William Crabtree also saw the transit of Venus, although he was a bit less fortunate than Horrocks. All day, clouds had obscured the view from his darkened apartment. Not until just before four o'clock that cold afternoon did they dissipate low above the western winter horizon of Manchester, giving him a view of the black droplet on the left side of the solar surface. He could have recorded the observation, as the coolheaded young curate had done, by tracing it out on a piece of paper placed up against the white screen on which he had projected the telescopic image. But Crabtree became so hypnotized by the event that by the time he recovered his composure the clouds had returned to spoil the view.

Just what secrets would Horrocks's observations disclose? Aside from being able to assess more accurately the diameter of Venus by measuring the size of the crisp disk against the bright surface of the sun, knowing the elements of Venus's orbit essentially would yield the scale of that orbit, both its shape and its size in relation to the earth's orbit. Did the planets really move on circles, as the Greeks propounded, or had God fashioned a heavenly ellipse for us to deal with? Horrocks may well have seized upon the idea of utilizing information about the Venus transit to determine the sun's true distance from earth, or at least some historians of astronomy have claimed so.[19] His discovery lay in recognizing that, with clock and telescope, Venus would be observed to pass over different parts of the sun at different times as seen from various locations on earth. The data could yield a parallax or displacement angle between two observation stations here on earth separated by a known distance. Horrocks's valuable information ultimately led to an accurate scale model of the solar system with all its orbital elements, pretty much as we know it today. For a self-educated man barely out of his teens, living far from the European mainstream of scientific progress, these are remarkable achievements.

These early telescope-wielding astronomers were making their first thrust at sizing up the universe in a process that started in earnest with the ideological change late in the Renaissance. What began then as the opening up of the depths of space that could be penetrated by human imagination and insight continues today as we have increased the power of the eye a billionfold and stretched the tape from millions of miles to billions of light-years.

As far as we know, Crabtree and Horrocks were the only two people on the face of the earth to watch the bright white light of Venus become transformed into an inkblot on the disk of the sun that Sunday in November, even though the event was visible throughout Europe and America; it is doubtful the handful of Pilgrims and Puritans who lived here would have been motivated to look at it. A transit of Venus would not happen again for 122 years. And by then 176 astronomers at 117 stations across the world would engage in cooperative study. Scientific progress took on explosive proportions in that century.

Later our two heroes planned to meet for the first time to drink a toast to their mutual triumph and talk astronomy late into the night. In a letter that survives, Horrocks promises his friend that he will make the twenty-five-mile journey between their respective cities on the fourth of January, 1641, "if nothing unforeseen should occur."[20] Prophetic words they turned out to be, for on the day before his planned departure, suddenly and inexplicably, Horrocks died, barely in his twenty-second year. For those who enjoy historical hindsight, it is worth pointing out that Horrocks left the scene of Cromwellian England before Newton ever arrived, and that his notebooks suggest he had anticipated, a number of undertakings the great synthesizer would later accomplish.

The stories of Galileo's and Horrocks's observations of Venus illustrate how channeling light through the telescope was part of a great shift in the way we think about the planets. The sky was no longer a place filled with moving lights that affect our destinies and passions; rather, it became a vast domain populated by worlds like our own— with surfaces, atmospheres, mountains, cores, and mantles—indifferent, inanimate worlds that whirl around one another under a different rule of law, one of unseen powers that are part of a natural rather than a divine order.

DIVERGING LIGHTS

While Galileo took long walks through the streets of Padua and Venice contemplating astronomy, and Horrocks invaded the crisp night sky over Britain with his spyglass, European culture was just beginning to come into contact with societies across the vast ocean on another continent. Eyes on the opposite side of the globe also looked up, though without contraptions or devices attached to them, at the lights in the sky. As I showed in Chapter 2, they witnessed tropical sky phenomena that were different, but, more important, they put their scientific knowledge to quite different uses. We know nothing of their scientific heroes, but what little we do know of their science leaves us wondering: Is the revolutionary foundation of our science unique? Did others too experience tensions between tradition and innovation? Did they unite the familiar with the unfamiliar to create new kinds of order?

These questions can be addressed by looking at other planet watchers at a time when there seems to be a clear indication that the tenor of the dialogue between people and nature was undergoing change. Before Columbus, the Americans were hermetically sealed from the European world by two oceans. Many cultures thrived there, each with a history that dated back to long before the fall of Rome. The first European contact with the mainland culture in America happened only two generations before Galileo. Born in the same year as Galileo, Shakespeare expresses his wonderment at the newfound civilizations on the other side of the world. Miranda's lines in Act V, Scene 1 of *The Tempest* describe her "vision of the island" (at once Naples and America):

O Wonder!
How many goodly creatures are there here!
How beauteous mankind is! O, brave new world,
That has such people in't.

Prospero answers: " 'Tis new to thee."

As we have seen in Chapter 4, New World astronomy had culminated with the Maya, more than a half millennium earlier. At a time when Europeans wrote in cumbersome Roman numerals, these people computed with dots and bars into the millions, making full use of the zero—and they were great sculptors, painters, and architects as well. Their "sidereal messenger" was the codex and their means of beaming light the ceremonial temple.

Using architecture as a means of enlightening humankind about the workings of nature? Although it may defy our common sense, the idea was prevalent for aeons in Europe even as it was in the New World. Consider the perspective of a pilgrim who entered the great portal of the cathedral of Chartres or Rouen, Cologne or Amiens. He saw every month of the year as a scene in a play about the human struggle with nature. Sowing, followed by reaping, then harvesting—all are portrayed in sculpted figures over the portal that represent the months by labors and the constellations of the zodiac. Next to them our pilgrim witnesses the seven sciences and the nine Muses, who teach the spirit of moral truth. Finally, as he enters the apse he is bathed in the light of God filtered through magnificent stained-glass windows upon which the history of creation is recorded in picture. Doorways and windows of the great medieval cathedral bind together the realities of nature and organized society for the common person.

The Mayan temple was no different. It, too, was an instrument that directed environmental imagery toward the sphere of religion rather than that of science. The temple was an information-laden house of knowledge, framed in a sacred environment for worship—an outdoor rather than indoor setting, for who would need to worship indoors in the tropics? But Mayan architecture may have done more for the believer than transport religious enlightenment through a window. In many cases Mayan ceremonial structures seem to have functioned as instruments deliberately positioned like an armada of telescopes to register the places of celestial bodies at the horizon for chronological purposes.

In the caracol of Chichén Itzá, an oddly oriented, highly assymetric building in north Yucatán that contains a peculiar central cylindrical element, Mayan astronomers built in four alignments that point to the eight-year Venus extremes. Two of these were erected through narrow horizontal shafts at the top of the turret. A third, perhaps even more fundamental, connected the exact center of the round core of the structure with the center of the stairway of its lower platform. In other words, the main axis of the building was aligned to Venus. Although no Venus symbols are found directly on the building, a set of five rain-god masks, with Venus symbols under the eyes, are situated on the northeast facade of a building that faces the caracol a hundred yards away. Around the corner on the lintel of the doorway to the east side of this building lies a representation of the Mayan zodiac, with Venus given star billing. In the middle of the skyband stands the Venus glyph flanked by various carved animal constellation: vulture-peccary-scorpion—the same ones that represent the constellations of the zodiac in the codices, each one perched upon that same Venus symbol.

Just a few days' walk through the brush to the west lie the ruins of Uxmal, and there, too, a planetary building is noticeably disoriented relative to the other buildings in its vicinity. We know little about the function of the so-called House of the Governor—whether it was a palatial residence or a public administrative center. We do know that, unlike all the others, it faces outward from Uxmal and looks off toward another Mayan city. A line perpendicular to its central doorway passes precisely through the only visible bump on the horizon—the largest structure at Cehtzuc, a neighboring center four miles away. The line also points in the approximate direction of one of Venus's eight-year extremes.

Over the Governor's doorways there is a frieze containing arrays of rain-god faces—each possessing Venus symbols under the eyes. And the masks are all arranged vertically in groups of five. Excavations on the northeast and more recently on the northwest corners of the structure's platform have revealed cornerstone masks with the number 8 appearing in dot-and-bar notation over the eyebrows. Was this number intended to indicate the eight-year cycle or perhaps the eight-day disappearance of Venus? We cannot really say, but I would favor the latter possibility because we know the temporal currency of the Maya seems always to have been the day rather than the year. The arrangement in fives and the connection between Venus appearances and the rainy seasons are also logical ties. (Recall that in Chapter 2 I showed that the duration of individual Venus disappearances actually changes with the seasons.)

In the caracol of Chichén Itzá and the House of the Governor at Uxmal, as well as in the Temple of Venus (Temple 22) at Copán, which was discussed in Chapter 4, there can be no doubt the Maya were keeping close watch on the planet Venus in order to determine the appropriate time to cast omens about the planting season and to celebrate a new metaphor between the accession of the young ruler and the sprouting of maize kernels that began the cycle of life—all of it celestially timed with unerring precision. The Spanish chronicler's description of a Maya astronomer (page 93) would fit Jeremiah Horrocks's job description perfectly. True, our astronomy has its centers and orbits, theirs a temporal rhythm making composed of the fours and fives of Venus's motion. Our astronomers, lens and notebook in hand, look up; theirs, parchment in hand, look through windows and doorways out to the horizon. Our scientists are abstract and esoteric, theirs concrete and agrarian; ours high-tech, theirs low-tech—almost no-tech. But in addition to diverging the celestial lights, let me single out a few points of convergence as well. Eighth-century Copán may not have been fifteenth-century Europe, but there are parallels.

We have some ideas about why Renaissance science developed an experimental philosophy toward nature and why explanations about the occurrence of natural phenomena came to be framed in a real world that people believed could be described mathematically and that operated via a cause-and-effect historical chain, quite separate from the theater of human consciousness. Historians believe that this Renaissance attitude arose out of major changes that had been taking place in West European culture, such as the gradual secularization of society, the spread of literacy, and the rise of an advanced technology that offered tangible ways of harnessing the natural world.

The great tension behind Renaissance scientific inquiry involved questioning the authority to rule. The charge was to explore the history of the ownership of authority and, consequently, the limits of free will. The great cultural rebirth was a political revolution, the scientific armature of which lay in the rediscovery, publication, probing, and questioning of those old Greek treatises on nature. These efforts revived the application to natural processes of mathematical and geometrical logic, which lies at the base of the description of the solar system by the use of space-bound orbital diagrams of the planets.

The empiricism that became wedded to mathematical logic was the unique contribution of the age. The drive to find the cold, hard facts of nature was a reaction to the long-entrenched philosophy that history's purpose was to set all humanity's course toward the forthcoming vision of the Apocalypse. No longer were scholars content simply to use the images of Venus to teach the dual love principle, or to explore her romance with Mars as a means of offering a lesson in moral behavior. If human action did not have to be directed toward some articulated future goal, then it could take place with significance for the sake of humanity here and now. The same would hold true for planets. In Western history the wanderers became the vehicles for a different inquiry: How do they move for themselves? And by what power, inherent or otherwise, do they describe their orbits?

This is *our* history. We need to ask the same kinds of questions of the eighth-century Maya. What motivated Mayan society to gravitate toward the *ahau* principle, the idea that the ruler of a dynasty was divinely and celestially ordained? Why did they suddenly look more closely at the timing of planetary events? What tensions had they experienced? Although it is fair to ask such questions, the answers, of course, are much more difficult to get at. We have no libraries filled with original Mayan history books—just a handful of scarcely decipherable painted bark fragments that remain the great conflagration of the conquistadors. Furthermore, the indigenous American cultures that survive are still very foreign and distant from the population that invaded this continent. Even today most people who reside in North and South America have little comprehension of the claimed sovereignty of the original native population.

Epigraphers who have managed to crack part of the hieroglyphic code and archaeologists who have substantially unveiled the stratigraphic layers of human history at some of the major Mayan sites are beginning to agree that the intense focus on what we would term esoteric matters, as in Renaissance Europe, occurred at a point in Mayan history when people would have had reason to question authority. It was then that they turned to a deeper investigation of the power inherent in the sky. For example, by the time of King Yax Pac, the sixteenth of his dynasty to rule Copán, the valley had become widely settled and the site had expanded outward from a small core settlement of a few thousand people to an overpopulated city of twenty thousand. Along with growth archaeologists have documented decay—soil erosion from cutting down trees, accompanied by a decline in rainfall and the spread of disease.

Overdeveloped and too heavily peopled for its base of subsistence, eighth-century Copán society clearly began to lose the ability to sustain itself economically.[21] It is only under these conditions that we can begin to understand Yax Pac's increased environmental curiosity and his ultimate appeal to the planetary gods that we explored in Chapter 4 as part of a fervent attempt to identify his rule with the Venus event that had ushered in the authority of his father. King Yax Pac's efforts to seek credibility for his divine rulership constituted a way of building upon and intensifying the agrarian notion of the Venus-Maize-Rain triad first exploited by 18 Rabbit, Yax Pac's predecessor by three generations removed and so exquisitely expressed in the architecture and sculpture of Temple 22. In effect, his reign signaled a small-scale Age of Empiricism in Copán.

As we have seen in Chapter 4, Yax Pac would tie most of the historical statements in the Copán inscriptions that pertained to his reign to the movements of the planet Venus, particularly its occurrence as evening star. His accession, the conduct of war, everything he says he did fell under the sign of his patron planet. Yax Pac altered history and guided his action by what he saw in the heavens.

On both sides of the Atlantic our stories are about changing ideology in the face of tension. Whereas the great intellectual revolution in which

Galileo and Horrocks participated in seventeenth-century Europe used the light of Venus to give the world a new, more rational, quantitative view of how the universe operates, an entirely different kind of revolution had happened in America before Columbus—a revolution we cannot fully appreciate because its background, aims, and goals are not even a remote part of our historical and cultural upbringing. The great Mayan innovation lay in the realization that celestial order is the ultimate source of all worldly things—that the ideal polity is the one contrived to function in perfect harmony with phenomena ordained by nature. And that is why Yax Pac's astronomers, whoever they were, wrote the great events of history in the immutable and precise currency of celestial events.

We have no license to pronounce the ancient Maya either right or wrong for how they fashioned their sky, for the irreducible facts of how the planets "really" behave can never be separated from the way of thinking, part religious and part scientific, that was embedded in Mayan culture. Surprisingly like our own ancestors before Copernicus and Galileo, the Maya used the sky as a system for managing their affairs, as a medium for a great dialogue on human-centered issues.

We can ask: Why didn't the Maya produce a Copernicus who recognized that the sun was the center of the solar system? Or that Venus, like the earth, moved in a planetary orbit in a space-bound universe of unlimited extent? But then, if they could speak to us, they might ask: Why didn't the Europeans pursue the obvious ties that we know exist between the cycles of nature and those of human affairs? Why didn't they make more of the fact that bright Venus —unlike Mars—disappears and reappears, always remaining close to the sun? How could Christians fail to see Venus as a way to express the resurrection principle, which forms such a basic part of their faith? And, they might wonder, why did their naked-eye sky watchers pick up on the quintessential Venus movements we dicussed in Chapter 2 while Galileo and Horrocks, careful observers though they were, failed to do so?

A famous epigrapher once wrote, "We must face it: so far as ends are concerned Maya astronomy is astrology."[22] As the codices and architecture demonstrate, the desire to penetrate nature in a more precise and quantitative way need not be motivated by a philosophy that detaches matter from spirit and reduces the former to a universal set of abstract laws. And a society's state of technology need neither stand in the way of promoting new motives for scientific research nor prohibit the creation of novel uses for precise knowledge.

To regard all the cultures of the world other than our own as simply "traditional" is to imply without justification that for them there are no innovations, no doubters of the status quo, no forces of change motivated by social stresses and tensions. I am convinced that the court astronomers and architects of King Yax Pac of Copán were technical innovators every bit as motivated and filled with resolve as our own Galileo and Copernicus. What is missing for these nameless architects of change is our inability to secure their places in the Mayan version of history, whatever that may be, the way Einstein has secured Galileo's place in ours: "A man is here revealed who possesses the passionate will, the intelligence, and the courage to stand up as the representative of rational thinking against the host of those who, relying on the ignorance of the people and the indolence of teachers in priest's and scholar's garb, maintain and defend their positions of authority."[23]

CHAPTER 7

SCIENCE:

THE IMAGE FOR ITS OWN SAKE

> Science is not merely the outcome of instinctive faith.
> It also requires an active interest in the simple occur-
> rences of life for their own sake.
> —ALFRED NORTH WHITEHEAD, 1925, P 20.

MODELING THE DEITY

We have followed war and love goddess first with naked eyes, then with spyglass, now with Mariner, Venera, and Magellan probes. Today we stand ready to walk over the once-sacred bodies of our planetary deities.

I began by listening in on conversations between people and planets described in clay tablet and bark paper. The dialogue was always about *us*—human perceptions and emotions. The common sense of the times endorsed the logical attachment between human and celestial affairs—each part of a single, whole, and animate universe. But Jupiter no longer has the power to make us jovial, jolly, merry, and gay; Mercury mercurial, lighthearted, fickle, changeable, active, or flighty; nor Venus venereous, lustful, libidinous, or amorous.

We still engage the planets; however, we no longer feel the need to talk to them—much less to heed their advice. As British philosopher Alfred North Whitehead states it, our common sense now requires that the planets exist for their own sake. What they do has no direct bearing on human consciousness. What ancient manuscripts once described as sacred objects of veneration, our written record now portrays as physical bodies in an inanimate universe, objects to dissect down to their most basic component parts. For we are the distant grandchildren of Sir Francis Bacon, and we still burn with ambition to exactify, quantify, and mathematicize nature, to chip away at her facets so that each might be fitted into the appropriate compartment in the complex celestial taxonomy that underlies the new hierarchical order we have imposed upon the universe.

What does the love goddess look like close up and where does she fit into contemporary ideas about nature? Behind the telescope and at the controls of the spacecraft resides a cadre of planetary explorers who ask other questions and seek other answers—predictions and retrodictions that our ancestors never could have dreamed up. Space scientists argue that, by studying the atmosphere and surface of Venus, we will be able to know more about the evolution of the solar system—where we came from and what the future holds. Like the Mayan planet watchers who established the birth of the gods at Palenque, astronomers today are engaged in tracing our beginnings—not just that of our species but that of every form of matter, animate or inanimate, in the universe.

It is no coincidence that at least since the time of Charles Darwin, scientists who study the solar system have been preoccupied with the question of how the sun and its retinue have *evolved*. Writing in the age of *The Origin of Species*, explorer-naturalist Alexander von Humboldt tells us in the volume on the planets in *Cosmos*, his massive work on the exploration of nature, that his desire is to show "how, by means of existing things, a small part of their genetic history is laid open."[1] He seeks to achieve his goal only by "the discovery of empirical laws," which give "insight into the causal connection of phenomena."[1] Like Darwin, Humboldt believed the key to understanding the past lay in a careful examination of what is tangible to us in the present, whether it be a page of historical writing, the archeological ruins of an ancient ceremonial center, the fossil remains of a Precambrian trilobite, or the image of the present condition of a planet as revealed through a telescope. Understanding the underlying causes of change remains squarely on the agenda of every scientific discipline. What caused the continents to overturn, the dinosaurs to disappear, the Black Plague to materialize? Our goal seems to be to acquire a chronological picture of change, a time-ordered sequence chain-linked by causes and effects—all arrived at through experience.

We inherited our practical way of thinking about the natural environment from sixth-century B.C. Milesian and Ionian philosophers such as Thales and Anaximenes, who, despite the label classical scholars assign them, were interested in how the universe works in the light of everyday experience. Because they were pragmatic thinkers, we might well expect them to have conveyed to their followers a conception of the universe above that works on the same familiar principles that guide the operation of those bits and pieces of it here below—the parts that are under our control. For example, the earth resides at the center of the universe. It acquires its stability because it floats on water. As such, earth supports and gives life to everything with which it interacts. We witness this in the many forms of life that thrive in a woodland pond, in rain that falls from the sky, in springs that rush out of the ground. For these natural philosophers, Greece was a living peninsula floating in a rich, beautiful, nurturing sea.

Where, then, did the planets come from and how did our earth become one of the diverse brood that follow mother sun across the busy road of the Milky Way? One theory proposes that the planets formed just as the sun did, condensing by the power of mutual gravitational forces out of a nebula that contained the same atomic composition as a typical gas cloud, with a small admixture of grainy material, located in the disk of our galaxy. The recipe consists mostly of hydrogen and helium, with added carbon, nitrogen, and a small percentage of heavier elements left over from earlier generations of stars that exploded and enriched the interstellar medium. The atmospheres of the planets were then formed by gases released from the molten planetary interiors. Next, chemical reactions and evaporation resulted from solar heating. Each of these processes was highly dependent on a given planet's distance from the sun's heat; this is one reason why the atmospheres of the planets have evolved to their present remarkably diverse conditions. Begun 4½ billion years ago, these evolutionary processes took at least a few hundred million years, by which time most planets acquired their own characteristic faces—the one we see close up in Mariner, Voyager, and Magellan video imagery.

A variation on this solar nebula theory proposes that the planets started not as contracting balls of gas that lost mass but instead as small grains of matter that accreted, like sticky snowballs rolling downhill. Once the newly formed sun began to heat these grainy masses, their volatile materials boiled off into enveloping shrouds, where chemistry and evaporation enhanced or reduced the atmospheres to yield what we observe today.

Still a third, overlapping chronology hypothesizes that, once the planets were formed, they were pelted by an invasive storm of material produced by effects outside the solar system—perhaps a nearby supernova explosion—which radically altered the planets' composition.

I am not concerned to venture into a detailed discussion of the myriad ramifications of each of these theories. The important point is that what drives our inquiry in formulating such explanations is the desire to know the intricate details of the rational set of processes that led to the universe as we know it today. Not only do we feel the need to know precisely where we came from but we also require that our present condition be expressed in the form of a history that charts out the detailed sequence of related events and finds in the seeds of one the outcome of the next. We validate this knowledge by physically intervening, testing, and analyzing those portions of the material record that can be made accessible to the senses. We need to see for ourselves whether the results of our tests support the idea or theory we propose. What is reasonable must not stray too far from our norms of common sense. If the tests fail, we must alter the idea and make further tests. In this creed we place our faith.

Earth and Venus look so much alike they could pass for sisters, at least as seen from a distance. They are almost the same size, the same mass, and the same density. They even orbit the sun at the same ballpark distance. But forget any close facial resemblance. We look with wonder and disbelief at the paradox we see through the eyes of Venera and Magellan: a fiery hell with a 900-degree surface temperature, the hottest place in the solar system. Venusian rock continents float in a stiff, plastic mantle of magma, a parched, waterless world deprived of oxygen yet with a carbon dioxide atmosphere so thick that a twenty-mile-an-hour wind could knock over any hypothetical tall buildings our minds might erect upon its surface. In an uncanny way, the personality of Venus that emerges seems remarkably close to that of old Babylonian Ishtar the morning after.

To what do we attribute the differences between lush, green earth and her wayward sister? Is Venus just barely close enough to the sun, compared with earth, for all its water to have evaporated and a runaway greenhouse effect to have developed? Great volumes of Venus's carbon dioxide were likely released from its surface rocks, whereas most of the earth's share of that gas still lies imprisoned in our relatively cooler environment. Just as water vapor clouds on earth can keep cloudy nights a bit warmer than clear ones, Venus's carbon dioxide absorbs infrared radiation that would otherwise leave the planet. As energy continues to be received from the sun, the temperature below the cloud layer is driven upward. Consequently, much of our current Venus probing concentrates on examining the content of the atmosphere and in particular the nature of the scant water molecules that can be observed there. We hope that gaining this knowledge will contribute to an understanding of how planetary greenhouses (including the one here on earth) operate.

From afar, Mars looks like earth, too: same day length, same tilt of its axis of rotation to the plane of its orbit, although its years and seasons are about twice as long. When first penetrated with powerful telescopes a century ago, the fiery red God of War was resolved to a blend of ocher deserts dotted with oases. Amateur astronomer Percival Lowell thought he saw verdant continents and nurturing lakes joined by waterways. He even charted canals that he imagined had been built by a superintelligent race of giants who had achieved peace in the face of a global Martian water shortage. Only by international cooperation to develop planetwide irrigation systems could they save themselves. Interesting that Lowell authored his imaginative Mars books just prior to the decade that spawned the League of Nations.

The problem with Mars, like that of Venus, is that when we dissect and test its properties close up, we no longer look into a mirror. It, too, becomes a unique world, like neither the moon nor Venus—clearly not of this earth. For example, Mars's polar ice caps, which Lowell witnessed shrinking and expanding seasonally, are made not of water, like our own, but of thermally trapped carbon dioxide. There is not enough water on Mars's surface to fill Lake Michigan. This fact alone is enough to render the planet lifeless, at least on the surface and in the present epoch, a conclusion confirmed by the Mariner surface landing almost a generation ago.

We may have set our eyes upon the Martian surface, but we cannot completely understand the planet from a single landing and a series of snapshots. We have cast the die of direct contact, and now our strategy commits us to more detailed exploration. We have left all but one or two stones on Mars unturned to date, but soon we will begin to get to the heart of it. Argues science journalist John Noble Wilford: "If Mars is not a central component of [a long-term] space strategy, Americans may forfeit their place in the vanguard of the human future that will be lived outside the cradle of Earth."[2] For Wilford, as for planetologists, the issue is not whether, only when, we will land on Mars and begin the process of "terrafirming" it—changing it into an earthlike province and availing ourselves of its material resources, perhaps over the protests of pro-Mars environmentalists. We treat planets the way some historians treat our ancestors: If at first .they seem unlike us, then we must restructure their images so that they resemble us more closely.

Are the Mars and Venus uncovered by modern technology deviants in a planetary madhouse, or is the earth the real victim of oddball evolution that has led us rather than our neighbors down a diverse historical path? Most planetary scientists now agree that a great catastrophe is the only way to account for the presence of the moon in our sky. A collision between the earth and a smaller mass early in the earth's history likely splashed a sizable droplet of planetary stuff into orbit around us, thus creating our silvery companion. The reheating generated in the interaction may have been appreciable enough to boil off and chemically alter the earth's atmosphere to a significant degree. This accident would account for why our air is only a fraction as dense as the Venusian shroud, which may emerge as the normal sort of atmosphere for a terra-sized planet at our distance from the sun. Some lunar geologists also have resorted to catastrophe to explain the composition of lunar rocks.

As odd as our planetary neighbors might appear, when we focus in on the planets farther out from the sun, we become aware of the existence of worlds more bizarre than any science fiction writer could have imagined. We have snooped on them, too, dropping probes that ricochet from one planetary host to another, unmasking them garment by garment. We say that we "experience" their environments through the data they relay back to us—data transformed into color-enhanced images and framed on a television monitor. From nearly 400 million miles away, Jupiter, just as the Greeks had indicated, emerges as the king of the gods. Largest of the planets, it (we can no longer say *he*) is enveloped in a colorful haze of intermingling yellows, reds, and browns. But all we really see is the skin on Jove's face—the top of an atmosphere that goes deeper than the width of our world and is so dense that, were you to stand at the bottom of it, you would be squashed to the thickness of a twenty-five-cent piece. We can safely analyze from afar the gases of the Jovian atmosphere, which we find to be composed of ammonia and methane laced with exotic hydrocarbons dissolved in an air of hydrogen—a lethal concoction that would poison our lungs were they not already crushed by gravity.

As I sit in an air-conditioned room looking at the Voyager photos, can I really say I have "experienced" Jupiter? In our modern world the value of a scientific idea is determined by its ability to predict events that take place in the sensate world. But you no longer use your nose to sniff the world directly, and what you predict goes well beyond the length of the nose technology has bequeathed to you. For example, planetary astronomers believe Jupiter's invisible interior consists of a small rocky core overlaid by liquid metallic hydrogen supporting an ocean of liquid hydrogen and an atmosphere of methane and ammonia. Allowing the universal gravitational and thermodynamic laws of nature to work on such a composition and distribution, they try to construct the model interior that best predicts the detailed characteristics of the ammonia snows, methane gases, and other atmospheric components that constitute the visible face of the planet.

Postulating a Jovian interior that consists of 60 percent hydrogen appeals further to our appreciation of a sense of unity in nature, for we know that the bulk of the universe is made up of hydrogen and only a little of this volatile substance could have escaped from the relatively cool portion of the original gas cloud that made up the solar system at the large distance of Jupiter from the sun. Proposing hydrogen in the high-pressure metallic form in Jupiter's core opens the gateway to the prediction that electric currents exist in the interior, and these in turn can be held responsible for conducting heat to the surface and generating a magnetic field into the space surrounding the planet, both of which we can observe. The marriage of theory and observation is elegant, even worth worshipping—when it works!

In a bizarre sense, the planetary gods, in the form of Jupiter and Saturn, who convened over Palenque to celebrate the archaic anniversary of the birth of the world were revealing their true faces on that warm July night in the tropics 1300 years ago, for what modern astronomers see of Jupiter and Saturn, more than Venus or Mars, comes closest to representing conditions as they were at about the time the solar system was born—atmospheres consisting mostly of hydrogen and helium, the most abundant original elements in the proto-solar system. Their atmospheres have remained chemically undifferentiated because these planets are so remote from the solar heat source. They are the solar system's great museum of planetary antiquity.

Since Apollo's man-on-the-moon mission, our dialogue with the planets has opened into a prolonged period of data acquisition directed toward building better planetary models. We live in a time not unlike early seventeenth-century Europe, when the optic reed became the technological innovation of a new planetary age, but with one significant difference: Our age is less rife with revolutionary ideas than with new technical ways of acquiring data. Today planetary theory lags far behind planetary observation.

If we stop and think about it, the models we create are a little like the machines we have grown to depend upon. Gears constitute the chain of reasoning; the parts are the assumptions; the function is the generation of explanation by mathematical law. When we say it works, we mean that the model cranks out accurate predictions that can be tested with real machines in the real world. Prediction is the product of all good scientific thought machines.

Since the midseventeenth century, a machine age has conditioned us to think in these mechanistic terms. For French philosopher and mathematician René Descartes, all the universe was a machine that ran like a clock. Once God wound it up and set it into motion, it would run forever on its own self-evident mathematical principles. *How*, not *why* the machine works has been science's main concern, and we have tinkered with its parts ever since that Cartesian revelation.

Like good machines, our most satisfactory planetary models are constructed in a minimalist way; that is, they are fitted with the least number of troublesome parts. For the sake of efficiency when they do not work, they must be capable of being stripped down into their interconnected components, and those parts that do not function well must be altered or replaced with new ones.

Likeness begets likeness as we fabricate each model out of a supply house of already familiar ideas. How else could we understand what we do not know? We use a billiard ball analogy to describe Voyager's trips through the solar system, and we invent a solar system model of the atom. We liken DNA to a computer resembling a ladder, with genetic materials that plug into terminals. We map the human genome like an underground subway system, and the nervous system becomes an electrical network along which chemically coded messages flow—all likenesses a culture such as ours would be expected to use as the grist for its scientific mill of metaphor.

A COMMON GROUND
WITH OUR ANCESTORS

have sketched out only the bare profiles of a few contemporary planetary faces. Many other texts are far more suitable for filling in the nuances and details. My concern is with what motivates diverse human societies to make planetary inquiries.

How can we even begin to place side by side the images of the planets we have in our minds today and those of our ancestors—to say nothing of those of foreign cultures who bear no connection to our past? The course of the modern West has been clear. Once Copernicus conceived of the idea of a moving earth, our personal stake in how the planets move was accompanied by a different set of concerns. Although Galileo's telescope brought unity to the wandering lights by envisioning each as a member of a solar family, the peering manmade eye that brought them up close began to reveal each sheep escaping from the fold of unified description.

From Galileo onward, we adopted a worldview that seeks to look upon the planets *as they really are*, denizens of a universe our culture invents in spatial terms, laden with matter and energy, and—especially since Darwin—a universe that had a definite beginning that is knowable and a changing, intelligible course that can be charted all the way to its predictable end. Our belief system confidently characterizes this evolving matter-and-energy universe as the only conceivable one, the one that exists for its sake alone. We need admit no other, past or present.

Philosopher Alfred North Whitehead has traced this faith in the existence of a natural chain-of-event order to the medieval belief in the existence of a rational God "conceived as with the personal energy of Jehovah," whose very word was an act of creation, "and with the rationality of a Greek philosopher"[3] Renaissance empiricists only needed to remove from the agenda the godliness of the planets and with it any reason for the wanderers to behave with due cause for the human conscience. They cut the astrological heart out of the astronomical body and offered up the world of nature for us to explain in pure historical terms. Now we insist that chronologically connected development be part of the way we know the natural world. It is the only way that satisfies us.

This evolutionary formula—of seeing the world around us in terms of change as opposed to status quo—permeates every aspect of society, including our belief in scientific and technological progress. It justifies chastising our ancestors for failing to have sought the underlying mechanisms that explain how things really work and crediting them only with having laid the empirical foundations of modern science. It allows us to deny them their cultural due. As far as we are concerned, there never was an Ishtar or a Quetzalcoatl. The universe each of them inhabited, the one our distant kin spoke and listened to, is all smoke and mirrors.

Yet I think there is common ground between us and them. Although we label ourselves scientific rationalists and our predecessors mystical pseudoscientists, we all share the dual belief that there is unity in the world and that we can experience it. Our ancestors sought unity between the behavior of the celestial gods and the seemingly chaotic affairs of people here below. They had faith that what they saw in the sky could help give structure to the society in which they lived. We, by contrast, believe in the physical unity of the structure of matter and energy and the natural laws that govern the behavior of the universe. The great spiral galaxy in Andromeda may be 2 million light-years away, but we are united with it in at least two respects. First, the gravity that it exerts on its satellite companion galaxy visible in the telescope is knowable by the same natural law that governs the mutual pull the earth and the moon or the earth and a falling apple exert upon each other. Second, the structure of the matter that makes up our extragalactic neighbor is basically no different from that of the stuff of which you and I are made. Atoms there, like those in our vicinity, consist of positively charged nuclei surrounded by clouds of negatively charged electrons; they, too, are bound by electromagnetic forces.

Gone are the exotic essences and emanations that once served to differentiate celestial from terrestrial matter. Even as late as the present century the element coronium, alleged to exist only in the sun, was proposed by some astrophysicists to account for the peculiar spectral lines in the solar corona witnessed during total eclipses. Nebulium, said to be present only in certain interstellar clouds, was hypothesized as a way of accounting for certain features in their electromagnetic spectra. These extraordinary localized substances, like Aristotle's heavenly quintessence, the divine element of which he supposed only sky-stuff was made, have been banished from existence and replaced by a pervasive, universal atomic and subatomic structure catalogued in the periodic table of the chemical elements. Even when we resolve the composition of the invisible, cold, dark matter that troubles many of today's cosmologists, I am confident that its structure will turn out to be familiar rather than exotic—for we are in the business of reducing the exotic to the commonplace.

Although we have not yet managed to amalgamate gravitational and electromagnetic forces under a single umbrella that can account for both the macroscopic and the microscopic behavior of matter and energy, we persist in the faith that this ultimate oneness will be found. As physicist-philosopher David Bohm writes, "Wholeness and unity are what is real. Fragmentation is the response of the whole to human action."[4]

We scarcely need to open up a popular scientific book or magazine to realize that it has become an article of faith that the universe is intelligible to us and that it can be comprehended rationally by extending things and processes we can understand on a practical, everyday basis to realms that, with less than a penetrating glance, seem perplexing. We are still questing after those hidden likenesses Jacob Bronowski talked about.

But there are planetary perspectives other than the scientific one, and they cannot be passed off via our own peculiar historical hindsight as merely superstitious. Far out of the realm of our common sense lie other invented universes powered by entities who enter into sexual union, entertain petty squabbles, and take out their frustrations on the human race. Our planets are propelled by neither whim nor decree of sovereign deities. Earlier I showed how both logic and experience have been integral parts of now-antiquated explanations about why the planets behave as they do. The basis of ancient Babylonian and Mayan planetary belief systems is different from our own in the sense that it rests upon a broader kind of faith: that the everyday earthly realm can be paralleled with what goes on in the outer world that envelops us and that these two realms really function in harmony. The universe was a distinct whole, with all parts intimately and intricately laced together, each aspect influencing the others. Nature and culture were one.

But common sense does not materialize out of thin air. How we see the world is really based on subjective criteria, culturally determined and historically conditioned by those who have influenced us, some from the recent, others from the distant and very distant past. Our appreciation of these influences is a matter of reasoned taste. It changes like the fashion of the times. Unidentified flying objects observed by nineteenth-century sky watchers looked like big black carriage hearses with huge spoked wheels. Scarcely a generation ago many historians characterized the Middle Ages as a time of rather low intellectual development and members of ancient societies were regarded as primitives with little inclination to pay close attention to what went on around them.

When we say that we seek to understand the planets as they exist for their own sake, we mean according to the dictates today's norms of common sense allow as reasonable. Our minds are capable of comprehending the planets only as places where power derives from certain abstract, indifferent essences, such as mass, charge, and strangeness—terms used to describe the qualities of nonvisible subatomic particles that are universal and mathematically expressible.

That Old Greek pragmatism has never really left us. We still think of scientific theories as acts of pure speculation and contemplation exercised by logic. Although they were developed as such by later Greek philosophers, we devise and deploy them in the development of planetary models pretty much in the old tradition of practical philosophers like Thales. The tradition of careful observation coupled with sound theory through the element of predictability serves up the planets we desire and deserve today. We have set ourselves a demanding agenda. Our theories must be efficient by involving the least number of complicating assumptions. More demanding still, they must be capable of leading to testable predictions, yet when the predictions are borne out of observations, they must be done so with unerring accuracy. Although scientific rather than metaphysical forecasting is now *en vogue*, for the uninitiated vast majority there is truly an element of magic in it.

IS COMMON SENSE CHANGING?

Millennarianism is an old idea—it happens once every thousand years. Just as time's odometer is about to turn over, people begin to speculate on whether a new age is about to dawn. Most cultures of the world once believed that the universe was destroyed and re-created on a grand time scale that repeated itself. The last great temporal stress point in the Western world happened late in the tenth century, when the Apocalypse was prophesied—the climactic Second Coming of Christ and his retinue of saints to the earthly domain, the great event for which all good Christians had already too patiently waited.

At mid–nineteenth century, science lived in a dream, poised on the edge of believing that, in principle, it could determine all that could be known about the universe—the position of every material particle at every point in time, past, present, and future. Ours was still a Cartesian universe. Feelings of confidence ultimately gave way in the twentieth century to the realization that a portion of nature always will remain hidden from us, veiled in uncertainty and chaos. Quantum mechanics was the tar baby that mired us in the reality that the very act of measurement interferes with what we seek to measure. Today no one will argue that observer and object are inseparable in the subatomic world, but the mere suggestion that the same could be true on a larger scale is scientific heresy.

The messages of our new millennium sound a more secular ring. For example, in *The End of Nature*, author Bill McKibben argues that the human species and the rest of the natural terrestrial environment can no longer be considered as separate systems, so deep and complex has the interaction between them become. Just as many contemporary scientists might view their way of knowing nature as the ultimate refinement of a long series of failed attempts to comprehend the real world, political scientist Francis Fukuyama sees the entire human past as a tortuous progression toward liberal democracy of precisely the type that has evolved in the United States today. In *The End of History* he posits that with the triumph of democracy over communism all ideological conflict in the world has been terminated.

Is science, too, at an end? Forces attempting to motivate currents of change in the conduct of science gather strength out of what they view as a conflict between the methods and goals of pure science and the interests of society in a world beset by poverty, hunger, overpopulation, and environmental destruction. Science as we know it advocates the exploration of nature as a rational act in itself, conducted dispassionately and motivated by a desire for objective truth. It is as separate from social concerns as nature is from human consciousness. The social context is not supposed to matter; in fact, paradoxically, the further society's needs are removed from concrete purposive pursuits, the nobler, the more selfless and dispassionate is the act of exploration in the eye of the true believer.

Critics have condemned certain kinds of pure research as acts of naked individualism and ego building. They ask whether we can afford to feed the insatiable appetites of physicists who require billions of dollars for toys such as superconducting supercolliders or of astronomers who ask for hundreds of millions of dollars to put a telescope in space. Like the ancient astrologer-priests they accuse of psychologically dominating their clients, scientists have been vilified as an elite shamanistic corps who alone think they possess the power to comprehend.*

So goes the criticism. And with it comes the accusation that institutionalized science has imperialized creative thought by placing the pursuit of truth under a catechism of the kind most succinctly stated by Nobel physicist Sheldon Glashow:

> *We believe that the world is knowable, that there are simple rules governing the behavior of matter and the evolution of the universe. We affirm that there are eternal, objective, extrahistorical, socially neutral, external and universal truths and that the assemblage of these truths is what we call physical science. Natural laws can be discovered that are universal, invariable, inviolate, genderless and verifiable. They may be found by men or women or by mixed collaborations of any obscene proportions. Any intelligent alien anywhere would have come upon the same logical system as we have to explain the structure of protons and the nature of supernovas. This statement I cannot prove, this statement I cannot justify. This is my faith.* [5]

---

* *The Joy of Insight* and *The Mind of God* are book titles that might invite such a perspective.

What credo do nonbelievers offer as a substitute? The great millennial change many of them tender seems concerned not just with seeking a more context-based, more socially conscious science; it calls for nothing less than restructuring scientific methodology down to its reductionistic roots—even the method of experimentation has been called into question.

A new biology, for example, challenges interventionism and intrusive experimentation, long the normal practice in medicine. CAT scanning, computerized tomography, and sonography are among the new noninvasive techniques for diagnosing ailments within the human body. Devices that monitor the depth for burned human tissue or the fullness of one's bladder add to the proliferation of noninterventional technology.

Physicist Hans von Baeyer theorizes that such products, methods, and techniques of the new NDE (nondestructive evaluation) are embedded in a spirit of change built around the revival of passive contemplation in science—the notion that there may be more to gain by exploring alternative ways of manipulating nature than by employing the Humpty-Dumpty method of taking things apart, studying each piece, and then trying to put them all back together again. Has the uncertainty principle finally taught us that intervention has its limits?

Millennial or not, the scientific debate is already in full swing, and the battle has been joined on a number of disciplinary fronts from medicine to cosmology, each focusing on humankind as the center of the universe. For example, although it has not gained much favor in the scientific community, the anthropic cosmological principle argues that the universe appears as we find it not because of some cosmic accident that took place 15 billion years ago, but instead because we are here.[6] Supporters argue that if the physical constants that make up the universe were much different from what we now measure them to be, then it would not be possible for life to exist. Given that a conscious observer requires a world in which to live, a spectator's very being becomes linked to all the universal constants. Not surprisingly, establishment scientists have widely assailed the theory as a mindless assault on cause-and-effect reasoning.

Where do the images of the planets fit in the millennial debates about change? James Lovelock's Gaia hypothesis offers us, in place of a material world apart, a living, pulsating organism that enjoins the development of life, human consciousness, and the evolution of the earth:

> *Gaia, as I see her, is no doting mother tolerant of misdemeanors, nor is she some fragile and delicate damsel in danger from brutal mankind. She is stern and tough, always keeping the world warm and comfortable for those who obey the rules, but ruthless in her destruction of those who transgress. Her unconscious goal is a planet fit for life. If humans stand in the way of*

this, we shall be eliminated with as little pity as would be shown by the micro-brain of an intercontinental ballistic nuclear missile in full flight to its target.[7]

This nurturing yet temperamental mother sounds akin to the one that Thales invented 2500 years ago and that Homer wrote about in the *Odyssey*, or Hesiod in his *Works and Days*, although both works were conceived over a hundred generations ago.

Gaia, named after the old Greek goddess, is the Earth returned to capital-letter status—no longer an inanimate spheroid eight thousand miles in diameter that just happens to be the stage that now supports a lucky agglomeration of successfully mutated forms. Gaia theory portrays earth and the life that it bears together as a single, interactive, self-maintaining system. Driven by solar energy, Gaia is capable of responding to changes in solar temperature, ocean salinity, and atmospheric composition forced upon it. Its ultimate subconscious goal is to perpetuate its capacity to regulate, to organize—to remain alive.

No one will question that environmental consciousness lies at the base of Gaia theorizing—just as social consciousness underlies the critique of scientific interventionism. Intended as mechanism rather than metaphor, it incorporates the idea that all the planets are not really the same insofar as the search for life is concerned. True, Venus may yet be "alive" with volcanic activity and Mars's canyons and streambeds once might have channeled water; but Mariner and Viking, Venera and Magellan delivered us the naked truth that brother Mars and sister Venus are barren of life. In Gaia-vision, life on earth is part of the control system that results in nitrogen and oxygen having become dominant in our atmosphere. These gases are almost totally absent on either Mars or Venus, whose atmospheres contain more than 95 percent carbon dioxide—largely waste gases in worlds that, like internal combustion engines, either have spent all their energy or were deprived of the energy needed to drive them in the first place.

The manned and unmanned space programs of the 1970s and '80s have demonstrated that humankind really is alone in the solar system. We now begin to face up to being the solitary inhabitants of the sun's entourage, if not of the entire universe. This is a far cry from the optimism of the 1960s, expressed for example in Carl Sagan and Iosif Shklovskii's *Intelligent Life in the Universe* and in the vain attempts of SETI (the Search for Extra-Terrestrial Intelligence) to reconnect us through radio communication to the rest of an environment we once felt confident throbbed with life.* No one then regarded us as an element of the causal end of the system. We were instead a part of the

---

* The U.S. Congress has recently authorized funds to revitalize SETI—perhaps part of a new wave of optimism.

great cosmic crapshoot, and an infinite number of rolls of the dice would be guaranteed to give us someone to talk to. In my view the feeling of solitude we now experience at the cosmic level is not so much a product of our present socially restrained and retrenched frame of reference as it is the outcome of scientific discovery. Cosmology texts of the sixties were written with a kind of naïve optimism that imagined not only rampant chemical and biological systems just like ours but also an intergalactic intelligentsia with the same cultural expansiveness, natural curiosity, and desire to explore—for the sake of acquiring knowledge itself—that earth-based ethnocentric theorizers possessed.

The demonstrable sensitivity of environment to humanity, the awareness of how special and precious life might be—these relatively recent revelations have produced new tensions that prompt earth scientists to think about the biosphere consisting not of coevolving organic and inorganic parts but rather as a single entity made up of living organisms: an atmosphere, an ocean, and a planetary surface all interacting with one another through a series of complex feedback processes. Housed in the same ideological casing as the anthropic cosmological principle, the Gaia scenario posits the climate and condition of planet earth not as a precondition to which life may or may not adapt but instead as the *result* of the way living organisms behave.

But how can life today be responsible for developments that happened yesterday? Aristotle once explained that a solid object falls to earth because of the necessity that, once displaced, it is compelled to reach its natural resting place together with all the other matter that makes up the element earth. Likewise, the Apocalypse was conceived as a forward pull along time's arrow that informs the practicing Christian that the morally good life that one ought to lead on earth today has its guidelines inscribed in a covenant between humanity and God cast in the future. Like the Christian Apocalypse and Aristotelian physics, Gaia has a strong teleological foundation or future-pulling force. It places cart before horse—offers ends to justify means and violates all our ideas of historical and chronological order. Its tensions have spawned a new ideology that tries to shatter the macroscopic arrow of time and its embedded cause-and-effect chain in the same way quantum physics attacked the dismantled once-sacred assumptions about the microscopic world.

Is the world more than the sum of its parts? Does it possess a purposive element? Is it conscious? Gaia seems to be part of a movement to rehitch culture to nature. It invites open conflict with Glashow's scientific catechism.

If we should reattach matter and spirit, what role might the planets play? We would then need to ask: Is Venus only a dead world, there for us to experiment with, to continue to probe for the sake of deriving information about how to improve our immediate environment, or might our sister planet, too, possess a planetary life history yet to be written, a story in which we have a part? Did she once have oceans and rivers like ours? How did the preconditions for life go awry? Why isn't she like us? These self-centered questions are no less remote than those our predecessors asked when they spoke to the planets just a few handfuls of generations ago. We cannot ignore them.

Or perhaps we perceive as dead what may be asleep? A University of Arizona group of planetologists has proposed, for example, that Mars undergoes a sleep-wake hydrological cycle hundreds of millions of years long.[8] Triggered by volcanic activity, cataclysmic outbursts of frozen underground water periodically flood the Martian surface. Released gases produce a greenhouse condition, resulting in a planetwide balmy climate. This is the scientists' way of explaining the contemporary landforms detailed by Viking photography: deep valleys and channels that once clearly carried water, permafrost, and terrains formed by the action of glaciers. Mars's depressed northern plains appear at one time to have been a vast ocean. Once water seeps back into the Martian turf and the sun dries out the land, the temperature falls and the planet enters its cold, dead phase once again, only to be awakened anew by internal forces.

Likewise, Voyager scientists had been puzzled to find the twin planets Uranus and Neptune exposing very different faces to the probing cameras. Uranus, quiescent and without an atmospheric blemish, appeared a calm contrast to violent Neptune, with its streaming clouds, howling upper-atmosphere winds, and turbulent oval splotch, a swirling hurricane as big as the earth. Could one planet be in the wake-up phase while the other lies dormant?

With the millennial odometer about to turn, the time-honored strategy of experiment and analysis, of intervention and manipulation seems, at least on the surface, secure against its counterpoint—contemplation and synthesis, nonintervention and passivity. But will science continue to command, and will nature continue to obey? That will depend upon how satisfied we remain with the outcome.

As the scientific doctrine of operating on a detached, observer-free universe persists and as technology proliferates with it, what can we predict? We are bound to witness the creation of scientific models that will offer us an ever more detailed understanding of the component pieces to which we continue to try to reduce the natural world—invisible exotica such as axions, neutralinos, WIMPs, MACHOs, cold dark matter, and quark nuggets. As we have divided the body in space into cell, cell into molecule, atom into nucleus and subnuclear particle, we are sure to penetrate beyond quark level in the near future. Likewise, we have stretched the chain of courses and events backward in time to cosmic ripples within a moment of God's creation. With some luck we may even develop overarching models that link gravity and quantum theory or biological symbiosis and mutation.

The scientific approach, as I have outlined it, has been for some time our key to knowledge, but will it unlock human destiny? For so long we have seen our fate written in Aristotle and Genesis. There is a principle of rule and subordination in nature at large that appears in the realm of animate creation, says Aristotle. This attitude of finding order through mastery over nature derives from a deeply rooted belief that hierarchy is a natural property of human existence—master over slave, parent over child, man over woman, human over animal, mind over body. Incidentally, as a teacher I have often argued, somewhat to the dismay of many curricular reformers, that one of the most effective ways to engage our students on the issues that matter to them is to trace with them the taproots of all our traditional beliefs, for better or for worse, through the great works of the past.

Division and hierarchical placement will always invite unrest among rational beings. And we who have placed ourselves at the top of the hierarchy know what we lack. What we need is a link between dialogues on the questions that are far more important to us and those with which everyday science deals as it gets increasingly out of touch with how we live our lives.

How do we age and die? Can we control the weather? Can we mimic intelligence? How does a cell become an embryo? For the layperson, these most oft-asked scientific questions of the day are only peripherally connected with the deeper social and moral inquiries that beset us. How we age has little to do with decisions about the correctness of prolonging life, and studying embryonic intelligence ultimately will not resolve the debate on abortion—nor will extending the cosmic yardstick help us decide whether we ought to mine the moon and Mars for natural resources. These are questions that, despite all the power we seem to be able to wield over nature with mature and skillful hand, we only feebly and ineptly entertain.

As cosmologist Steven Weinberg remarks in his popular book on the Big Bang theory of cosmology, *The First Three Minutes*, "The more the universe seems comprehensible, the more it also seems pointless. There is no solace in the fruits of our research."[9] Is there no other mandate for us but to be content working above our everyday concerns, writing the script of the great tragedy of universal evolution?

Will science ever be able to incorporate within its very soul a moral capacity? We may have criticized Greek pagan and Mayan astrology because they did not always work, but let us realize that these were enduring highly successful systems for managing human affairs and expressing human emotions. They framed the dialogue between people and their gods, between society and nature. What we regard as ancient dogmatic beliefs provided a measure of social satisfaction. Whether astrology did or did not work seems more our problem than that of the believers who took comfort in it for so many thousands of years.

Will we give back to nature the subjectivity and forethought we believe she once possessed in the eyes of our ancestors? Gaia evokes radical thought, and I have cast it as but one symptom of a kind of anthropocentric uprising that is attempting to take root in the science of the twenty-first century—an attitude sown of the seeds of dissatisfaction with the universe twentieth-century science has left us to ponder. Whether it will become a significant part of the five-millennia-old story of the interaction between planets and people, of the harmony and turbulence we have witnessed at the many interfaces between nature and culture, it is still too early to tell. In everyday life, most of us continue to think of the material and social spheres of the real world as quite distinct from each other. Mostly urban dwellers, only rarely do we use our senses to smell flowers in the field or look at the planets in a dark sky. As ordinary citizens we have fallen out of touch with nature. And as society becomes ever more wantonly unpredictable, I wonder whether we have not totally lost sight of nature as the only true measure of stability and constancy to follow—not to *lead* but to *follow*.

Science has strained, stressed, and contorted nature to see how it reacts—strictly for its own sake. But history and anthropology teach us that when people are dissatisfied with ideology they change it. Is a new dialogue about to begin?

EPILOGUE

Eavesdropping on the ageless dialogue between people and planets, I feel less inclined to highlight the diversity of what I have heard and spoken to you about. Rather I want to embrace even more firmly what I believe is at the root of the discovery process—seeking order and unity by finding in the unexplained something that we already know. I chose this idea as a common denominator for exploring the manifold ways of understanding the planets not because it is culturally democratic in an age of egalitarianism nor because relativism, as the epigraph states with a bit of sarcasm, is today's correct approach. I did so because it offers us a genuine antidote to one misguided alternative—the philosophy of progress, which openly discredits the past for lacking aspects of our modern perspective, an attitude that leaves us with a false feeling of superiority.

Like the tide, we have seen humanity's dialogue with the planets periodically advance and recede between precise observation and abstract symbolism. The astral mythologies of the European Middle Ages fled the world of the senses and developed an imaginative and elegant imagery of their own. By contrast, in early Greek and Classical Mayan society as well as in the contemporary mythology that we call science, there seems to have been a very close correspondence between what people saw and what they said, at least about the planets.

Yet in another sense people have not changed. As above, so below—we have always sought that parallel between what happens in nature and the events that affect our social and personal lives. We manifest our search in different ways: We make calendars to schedule festivals; we time the occurrence of celestial phenomena to find whatever accord might exist among other natural occurrences, such as seasonal changes, planting, the migration of birds; and social phenomena, such as the arrival of pestilence, war, and good or bad political fortune. So, Kukulcan heralded the dawn and prophesied war just as Venus became a convenient metaphor for resurrection. All were aspects of a future deemed foreseeable. And we share with our ancestors the undying belief that there really is order in this world, although we waffle more than they on its purposiveness.

In some stratified urban societies where segmented and specialized domains of knowledge developed, we saw the art of associating celestial sign and indication develop into a complex syncronic system. In certain instances, celestial signals were thought to perpetuate cause through form or emanation—the way Pliny suggested that low-lying Venus deposits her procreative dew upon the vegetation. Our tendency to dismiss such processes as superstitious mechanisms that convey false cause-and-effect relationships—even if they are the same kind as those embedded in the modern scientific definition of the real world—only serves to distance our past from ourselves. For the stories of the planets—their names, what they have represented, how those representations were altered and enhanced, the intricate details that emerge from the written form of the dialogue in documents such as the Venus Tablet of King Ammizaduga—offer us deep insights into the minds of those who have looked up to the sky before us, people who had no necessity to know nature in terms of mathematically based, universal, causal laws.

Instead of demystifying the flight of ancient Venus into the underworld, I have sought to show that such imagery is rooted, like our own science, in actual happenings reflected in the course the planet takes across the sky. Only the interpretation differs. But while the planetary perspectives I have sketched out occupy a common ground upon which we and ancient astronomers tread, we must not neglect one very significant difference: Modern Western culture willfully chose to separate nature from culture. It was a slow and delicate surgical separation, which began only twenty generations ago in the European Renaissance, and it is not yet complete, although most will concede that we all live in a scientific world of rational naturalism.

Lately we have been rethinking our thoughts about the planets, wondering whether the perspectives of the astrologers of old were really all that different from those of the scientific astronomers of today. We are well aware of the gains we've made with our invasive, tech-armed strategy for getting to the bottom of what makes nature tick. But we need to be aware of what we lost when we unglued the bond between matter and spirit and traded away the animated half of the world for the limited free will we have exercised so briefly to probe the other half of existence.

Today, Mariner and Magellan have carried us beyond the veil of the love goddess. We have depersonalized Mother Nature, convinced ourselves that the material world exists and behaves strictly in its own self-interest. But we are beginning to pay the price of living in a motherless world. We have become the outsiders, onlookers who can only helplessly watch the interactions among forces and entities that have neither care nor cause for us. We have become bewildered by a universe that we still want to regard as nurturing, yet one we see filled with so much violence, unpredictability, and chaos—a universe as fickle and uncaring as Ishtar the morning after. We struggle in our new covenant to control nature, and, although we find containment at the local level manageable, the big picture throws up overwhelming prospects we know we never can contend with. Cosmologists may think they can read God's palm, but they cannot keep the sun from evolving nor the universe from expanding.

In frustration, a few of us cling to newspaper astrology, a poor imitation of astronomy's cast-off stepmother. Others reach for exotica—UFOs and reincarnation. The more sober undertake scientific searches for intelligent life, but what could seem less satisfying than the realization that the self-regulating universe that indifferently envelops us has not somehow elsewhere spawned our kind of intellect, one that shares our indifferent inquisitiveness and desire to explore nature for nature's sake? How else can we account for our very consciousness?

What is the state of the planetary dialogue as we close this orbit? Our encounter with the subatomic world has revealed an environment in which the observer of nature is forced to become a participant in nature's play. At the global philosophical level, the Gaia theory and the anthropic cosmological principle constitute the latest attempts to restore actor-spectator interaction to the whole earth, to the solar system—even to the universe. We of the earth, like the Quiché Maya who received fire from Tohil, proclaim ourselves rightful heirs of the Big Bang. That such ideas are being developed at an acute period of environmental awareness is no accident. It only underlines the truth that science, despite its incandescent rise in the second half of the millennium, cannot completely free itself from culturally motivated attitudes and perspectives, can never remain truly socially neutral, in either method or application.

I would not dare predict—either with lens or horoscopic chart—whether the next century will witness an overhaul of scientific process and a restructuring of the links between human culture and the world of nature, but if twentieth-century science has revealed anything, it is that knowledge once regarded as absolute often emerges as transparent. It melts, disintegrates, and transforms again in our very hands, even as we try to focus in upon it to illuminate the unknown.

REFERENCES

Anderson, A. J. O., and Dibble, C, 1953, *Florentine Codex Book 7, The Sun, Moon and Stars, and the Binding of the Years*, Santa Fe: School of American Research and Salt Lake City: Universityof Utah.

Aristotle (see E. Barker).

Armitage, A. 1947, *The World of Copernicus*, New York: Mentor.

Augustine, St. (see R. Pine-Coffin).

Aveni, A. 1989, *Empires of Time*, New York: Basic.

Aveni, A. (ed.) 1992, *The Sky in Mayan Literature*, Oxford: Oxford Univ. Press.

Aveni, A. 1980, *Skywatchers of Ancient Mexico*, Austin: Univ. of Texas Press.

Bacon, F. (see. J. Weinberger).

Barker, E. 1958, *Aristotle: Politics*, London: Oxford Univ. Press.

Barnes, R. 1974, *Kédang: A Study of the Collective Thought of an Eastern Indonesian People*, Oxford: Clarendon.

Bartusiak, M. 1990, Review of J. Wilford's *Mars Beckons*, *New York Times Review of Books* 15 July, 1.

Beer, A., and P. Beer, ed. *Mysterium Cosmographicum*, in *Kepler, 400 Years*, New York: Pergamon.

Bloom, A. 1987, *The Closing of the American Mind*, New York: Simon & Schuster.

Bohm, D. 1980, *Wholeness and the Implicate Order*, London: Routledge & Kegan Paul.

Bok, B. 1975, "A Critical Look at Astrology," *The Humanist*, 35 (5), 6–9.

Bok, B. 1975, "Objections to Astrology," *The Humanist*, 35 (5), 4–5.

Bronowski, J. 1965, *Science and Human Values* (rev. ed.), New York: Harper & Row.

Buber, M. 1958, "The Wonder on the Sea," in M. Buber, *Moses: The Revelation and the Covenant*, New York: Harper & Row, 74–79.

Burgess, E. 1985, *Venus, An Errant Twin*, New York: Columbia Univ. Press.

Burkert, W. 1983, *Homo Necans, The Anthropology of Ancient Greek Sacrificial Ritual and Myth*, Berkeley: Univ. of California Press.

Campion, N. 1982, *An Introduction to the History of Astrology*, London: Inst. for the Study of Cycles in World Affairs.

Capra, F. 1982, *The Turning Point: Science, Society and the Rising Culture*, New York: Simon & Schuster.

Chamberlain, V. 1982, *When Stars Came Down to Earth: Cosmology of the Skidi Pawnee Indians of North America*, Los Altos: Ballena; College Park: Center for Archaeoastronomy.

REFERENCES

Chapman, A. 1990, "Jeremiah Horrocks, the Transit of Venus, and the 'New Astronomy' in Early 17th Century England," *Quarterly Journal of the Royal Astronomical Society* 31:333–357.

Clagett, M. 1989, *Ancient Egyptian Science*, (2 vols.) Philadelphia: American Philosophical Society.

Closs, M. n.d., "Extreme Point of Venus in Maya Writing" (ms 1987).

Closs, M., A. Aveni, and B. Crowley. 1984, "The Planet Venus and Temple 22 at Copán," *Indiana* 9:221–47.

Codex Cospi, Univ. of Bologna Library.

Codex Dresden, Royal Public Library, Dresden.

Copernicus, N. (see J. Dobson and S. Brodetsky)

Crump, T. 1990, *The Anthropology of Numbers*, Cambridge: Cambridge University Press.

Culver, R., and P. Ianna. 1979, *The Gemini Syndrome, Star Wars of the Oldest Kind*, Tucson: Pachart.

Curry, P. (ed.). 1987, *Astrology, Science and Society*, Wolfeboro, NH: Boydell.

Dante (see C. Singleton).

DeKosky, R. 1979, *Knowledge and Cosmos: Development and Decline*, Washington: University Press of America.

Dick, D. 1970, *Early Greek Astronomy to Aristotle*, London: Thames & Hudson.

Dobson, J., and S. Brodetsky. 1947, Nicolaus Copernicus, *De Revolutionibus Oribum Coelestium*, Preface and Book I. Royal Astronomical Society, London.

Drake, S. 1957, *Discoveries, and Opinions of Galileo*, Garden City, NY: Doubleday, Anchor.

Drake, S. 1967, *Galileo Galilei, Dialogue Concerning the Two Chief World Systems*, Berkeley: Univ. of California Press.

Dütting, D. 1982, "The 2 Cib 14 Mol Event in the Inscriptions of Palenque, Chiapas, Mexico," *Zeitschrift für Ethnologie* 107(2):233–58.

Falkenstein, A. 1936, *Archaische Texte aus Uruk*, Leipzig.

Fash, W., and B. Fash. 1990, "Scribes, Warriors and Kings: The Lives of the Copán Maya," *Archaeology* 43(3):26–36.

Field, J. V. 1987, "Astrology in Kepler's Cosmology," in *Astrology, Science and Society*, ed. P. Curry, Suffolk: Boydell.

Fraknoi, A. 1989, "Your Astrology Defense Kit," *Sky & Telescope* (Aug.): 146–50.

Frankfort, H. 1948, *Ancient Egyptian Religion: An Interpretation*, New York: Harper & Row.

Frazer, R. M. (tr. and ed.). 1983, *The Poems of Hesiod*, Norman: Univ. of Oklahoma Press.

Freidel, D., and L. Schele. 1988, "Kingship in the Late Preclassic Maya Lowlands: The Instruments and Places of Ritual Power," *American Anthropologist* 90:547–67.

Fulke, W. 1979, *Book of Meteors*, Mem. Am. Phil. Soc. 130.

Fukuyama, F. 1992, *The End of History and the Last Man*, New York: Free Press.

REFERENCES

Gale, G. 1981, "The Anthropic Principle," *Scientific American* 245(6):154–171.

Galileo (see S. Drake and see A. van Helden).

Garin, E. 1983, *Astronomy in the Renaissance, The Zodiac of Life*, London: Routledge & Kegan Paul.

Garland, R. 1990, *The Greek Way of Life*, London: Duckworth.

Gauquelin, M. 1967, *The Cosmic Clocks: From Astrology to Modern Science*, Chicago: Univ. of Chicago Press.

Gauquelin, M. 1973, *Cosmic Influence on Human Behavior*, New York: Stein & Day.

Gauquelin, M. 1979, *Dreams and Illusions of Astrology*, Buffalo: Prometheus.

Gauquelin, M. 1969, *The Night of Astrology*, New York: Stein & Day.

Gauquelin, M. 1969, *The Scientific Basis of Astrology*, New York: Stein & Day.

Gleadow, R. 1969, *The Origin of the Zodiac*, New York: Atheneum.

Goldsmith, D. (ed.). 1977, *Scientists Confront Velikovsky*, Ithaca: Cornell Univ. Press.

Gordon, R. 1977, "The Sacred Geography of a Mithraeum: The Example of Sette Sfere," *Journal of Mithraic Studies* 1(2):119–65.

Gordon, R. 1989, *The Origins of the Mithraic Mysteries*, Oxford: Oxford Univ. Press.

Harding, S., M. Hesse, S. Glashow, 1989, "Does Ideology Stop at the Laboratory Door?" *The New York Times*. 22 Oct., 24.

Hawkes, J. 1967, "God in the Machine," *Antiquity* 51:174–80.

Hawkes, J. 1962, *Man and the Sun*, New York: Random House.

Heiberg, J. L. (ed.). 1898, *Ptolemy, Syntaxis Mathematica*, Leipzig: Teubner.

Hesiod (see R. M. Frazer).

Homer (see E. Rieu).

Hostetter, 1990, "Naked Eye Crescent of Venus," *Sky & Telescope* (Jan.): 74–76.

Huber, P. 1977, "Early Cuneiform Evidence of Venus," in *Scientists Confront Velikovsky*, ed. D. Goldsmith, Ithaca: Cornell Univ. Press.

Humboldt, A. von. 1871, *Cosmos: A Sketch of the Physical Description of the Universe*, London: Bell and Daldy.

Jastrow, M. 1911, *Religious Belief in Babylonia and Assyria*, London: G. P. Putnam's.

Jerome, J. 1991, "Ancient Seas of Mars," *The Sciences* 31(6), 7.

Jerome, L. 1977, *Astrology Disproved*, Buffalo: Prometheus.

Jones, W. 1982, *Venus and Sothis*, Chicago: Nelson-Hall.

Kaula, W. 1990, "Venus: A Contrast in Evolution to Earth," *Science* 247:1191–96.

Kepler, J., (see A. Beer and P. Beer).

Kitson, A. 1984, *History and Astrology*, London: Unwin.

Koestler, A. 1960, *The Watershed*, New York: Doubleday.

Kramer, S. 1961, *Sumerian Mythology*, New York: Harper & Row.

Kuhn, T. 1957, *The Copernican Revolution*, New York: Vintage.

Landa, D. de (see A. Tozzer, ed.).

REFERENCES

Langdon, S., J. Fotheringham, and C. Schoch. 1928, *The Venus Tablets of Ammizaduga*, London: Humphrey Miller and Oxford Univ. Press.

Larios, R., and W. Fash. n.d., "Architectural History and Political Symbolism of Temple 22, Copán," paper presented at Palenque Mesa Redonda VII 1989 (in press).

Latham, R. (ed.). 1951, *Lucretius, On the Nature of the Universe*, New York: Penguin.

López Austin, A. 1974, in *Sixteenth Century Mexico*, ed. M. Edmonson, Albuquerque: Univ. of New Mexico Press, 111–49.

Lounsbury, F. 1983, "The Base of the Venus Table of the Dresden Codex and Its Significance for the Calendar Correlation Problem," in A. Aveni and G. Brotherston (eds.), *Calendars in Mesoamerica and Peru*, Oxford: British Archaeological Reports, International Series, 174:1–26.

Lovelock, J. 1988, *The Ages of Gaia, a Biography of Our Living Earth*, New York: W. W. Norton.

Lucretius (see R. Latham).

MacLeod, B. n.d. "The Crossing Triplet Gods: A Reading for the Triad Introductory Glyph," paper presented at Mesa Redonda VII Palenque Round Table, Palenque, 1989 (in press).

Makemson, M. 1941, *The Morning Star Rises: An Account of Polynesian Astronomy*, New Haven: Yale Univ. Press.

Mason, H. (ed.). 1972, *Gilgamesh, A Verse Narrative*, New York: Mentor.

Maspero, G. 1896, *The Dawn of Civilization*, London: Society for Promoting Christian Knowledge.

McCaffery, E. 1942, *Astrology, Its History and Influence in the Western World*, New York: Scribner's.

McKibben, W. 1989, *The End of Nature*. New York: Random House.

Melchiori, K. 1989, Shakespeare, *King Henry IV, Part 2*, Cambridge: Cambridge University Press.

Miller, M. 1988, "The Meaning and Function of the Main Acropolis, Copán," in *The S.E. Classic Maya Zone*, ed. E. Boone and G. Willey, Washington: Dumbarton Oaks.

Moore, P. 1959, *The Planet Venus*, New York: Macmillan.

Nathan, A. J. 1991, "Tiananmen and the Cosmos," *New Republic* 29 July, 31–36.

Neugebauer, O. 1983, "Mathematical Methods in Ancient Astronomy," in O. Neugebauer, *Astronomy and History, Selected Essays*, New York: Springer-Verlag.

Neugebauer, O. 1941, "Some Fundamental Concepts in Ancient Astronomy," in *Studies in the History of Science*, Philadelphia: Univ. of Pennsylvania.

Neugebauer, O. and R. Parker. 1969, *Egyptian Astronomical Texts III. Decans, Planets, Constellations, and Zodiacs*, Providence: Brown Univ. Press.

New, D., and J. New. 1962, "The Dances of Honeybees at Small Zenith Distances of the Sun," *Journal of Experimental Biology* 39:271–91.

REFERENCES

North, J. D. 1986, "Celestial Influence—The Major Premiss in Astrology," in *Astrologi Hallucinati*, ed. P. Zambelli, Berlin: deGruyter.

Nuttall, Z. 1904, "Periodical Adjustments of the Ancient Mexican Calendar," *American Anthropologist* 6(4):486–500.

Offord, J. 1915, "The Deity of the Crescent Venus in Ancient Western Asia," *Journal of the Royal Asiatic Society, Great Britain and Ireland* 26:1197.

Oppenheim, A. 1969, "Divination and Celestial Observation in the Last Assyrian Empire," *Centaurus* 14:97–135.

Pannekoek, A. 1961, *A History of Astronomy*, New York: Dover.

Panofsky, E. 1962, *Studies of Iconology: Humanistic Themes in the Renaissance*, New York: Harper & Row.

Parker, R. 1974, "Ancient Egyptian Astronomy," in *The Place of Astronomy in the Ancient World*, ed. F. Hodson, *Philosophical Transactions of the Royal Society*, (London), A(276):51–66.

Paz, O. 1976 in J. LaFaye, *Quetzalcoatl and Guadalupe: The Formation of Mexican National Consciousness, 1531–1813*, tr. B. Keen, Chicago: Univ. of Chicago Press.

Pine-Coffin, R. (ed.). 1961, *Saint Augustine, Confessions*, New York: Penguin.

Pingree, D. 1963, "Astronomy and Astrology in India and Iran," *Isis* 54: 229–46.

Pritchard, J. 1955, *Ancient Near Eastern Texts Relating to the Old Testament*, Princeton: Princeton Univ. Press.

Proctor, R. 1875, *Transits of Venus. A Popular Account of Past and Coming Transits, from the First Observed by Horrocks A.D. 1639 to the Transit of A.D. 2012*, New York: Worthington & Co.

Ptolemy (see J. L. Heiberg).

Reiner, E., and D. Pingree. 1975, *Babylonian Planetary Omens Part One, Enuma Anu Enlil Tablet. 63: The Venus Tablet of Ammizaduga*, Malibu: Undena.

Rieu, E. (tr.). 1950, *Homer, The Iliad*, New York: Penguin.

Righini, G., and T. Settle. 1976, "G. Alfonso Borelli e la Visibilitá di Venere," *Annuario del Institutio e Museo. di Storia della Scienza di Firenze* 1(2):37–56.

Roe, P. 1982, *The Cosmic Zygote: Cosmology in the Amazon Basin*, New Brunswick: Rutgers Univ. Press.

Roys, R. 1933, *The Book of Chilam Balam*, Washington: Carnegie Institution of Washington, Pub. 438.

Rykwert, J. 1986, *The Idea of a Town*, Princeton: Princeton Univ. Press.

Sachs, A. 1974, "Babylonian Observational Astronomy," in *The Place of Astronomy in the Ancient World*, ed. F. R. Hodson, London: Royal Society.

Sagan, C. and I. Shklovskii, 1966, *Intelligent Life in the Universe*, New York: Dell.

Sahagún, B. (see A. J. O. Anderson & C. Dibble, tr.).

Sarton, G. 1950, Review of E. Drover's *Book of the Zodiac*, *Isis* 41:374.

Sayce, A. 1874, "Astronomy and Astrology of the Babylonians," *Tr. Society of Biblical Archaeology* 3(1):145–339.

REFERENCES

Schafer, E. 1977, *Pacing the Void*, Berkeley: Univ. of California Press.

Schele, L., and M. Miller. 1986, *Blood of Kings*, Ft. Worth: Kimbell Museum.

Seler, E. 1904, *Annales de Quauhtitlan*, tr. and parenthetical remarks by E. Seler, *Bureau of American Ethnology Bulletin* 28:364–65.

Seznec, J. 1953, *The Survival of the Pagan Gods*, Princeton: Princeton Univ. Press.

Shakespeare, W., see G. Melchiori, ed.

Singleton, C. 1978, *Dante Alighieri, Divine Comedy*, Princeton: Princeton Univ. Press.

Sprajc, I. n.d. "Venus and Temple 22 at Copán Revisited," *Archaeoastronomy* Supplement to the *Journal for the History of Astronomy* (in press).

Tambiah, S. 1990, *Magic, Science, Religion, and the Scope of Rationality*, Cambridge: Cambridge Univ. Press.

Tedlock, B. 1982, *Time and the Highland Maya*, Albuquerque: University of New Mexico Press.

Tedlock, D. (tr. and ed.). 1985, *Popol Vuh: The Mayan Book of the Dawn of Life*, New York: Simon & Schuster.

Tester, S. J. 1987, *A History of Western Astrology*, Wolfeboro, NH: Boydell.

Thompson, J.E.S. 1960, *Maya Hieroglyphic Writing*, Norman: Univ. of Okla. Press.

Thompson, J.E.S. 1970, *Maya History and Religion*, Norman: Univ. of Okla. Press.

Thompson, J.E.S. 1972, *A Commentary on the Dresden Codex, A Maya Picture Book*, Philadelphia: American Philosophical Society.

Thompson, R. 1900, *Reports of the Magicians and Astrologers of Nineveh and Babylon in the British Museum*, 2, London: Luzac.

Thorndike, L. 1923, *A History of Magic and Experimental Science During the First Thirteen Centuries of our Era*, New York: Columbia Univ. Press.

Tozzer, A. 1941, "Landa's Relación de las Cosas de Yucatan," Cambridge, Mass.: Papers of the Peabody Museum of American Archaeology and Ethnology, Harvard Univ, vol. XVIII.

Toulmin, S., and J. Goodfield. 1965, *The Discovery of Time*, Chicago: Univ. of Chicago Press.

Toulmin, S., and J. Goodfield. 1961, *The Fabric of the Heavens*, New York: Harper & Row.

Truzzi, M. 1972, "The Occult Revival as Popular Culture: Some Observations on the Old and the Nouveau Witch," *Sociological Quarterly* 13:11–36.

Ulansey, D. 1989, "The Mithraic Mysteries," *Scientific American* 261(6): 130–35.

Ulansey, D. 1989, *The Origins of the Mithraic Mysteries*, Oxford: Oxford Univ. Press.

van der Waerden, B. 1974, *Science Awakening II, The Birth of Astronomy*, Leyden: Noordhoff; New York: Oxford Univ. Press.

van Helden, A. 1985, *Measuring the Universe*, Chicago: Univ. of Chicago Press.

REFERENCES

van Helden, A. (tr. and ed.). 1990, *Siderius Nuncius or the Sidereal Messenger, Galileo Galilei*, Chicago: Univ. of Chicago Press.

Velikovsky, I. 1950, *Worlds in Collision*, New York: Doubleday.

Vernant, J. P. 1983, *Myth and Thought Among the Greeks*, London: Routledge & Kegan Paul.

Villacorta, J, and C. Villacorta. 1977, *Codices Mayas*, Guatemala City: Tipografía Nacional.

von Baeyer, H. 1989, "A Commanding View," *The Sciences* 29(5):6–9.

Weber, R. 1986, *Dialogues with Scientists and Sages: The Search for Unity*, London: Routledge & Kegan Paul.

Weinberg, S. 1988, *The First Three Minutes*, New York: Basic.

Weinberger, J. (ed.). 1980, *The Great Instauration and the New Atlantis*, Arlington Heights, IL: Harlem Davidson.

Weir, J. 1972, "The Venus Tablets of Ammizaduga," Istanbul: Nederlands Historisch Archaeologisch Instituut in Het Nabije Oosten.

Whatton, A. 1859, *The Transit of Venus across the Sun, A Translation of the Celebrated Discourse thereupon by the Rev. J. Horrox*, London: Wm. MacIntosh.

Whitehead, A. N. 1925, *Science and the Modern World*, New York: Macmillan.

Wilcox, D. 1987, *The Measure of Times Past*, Chicago: Univ. of Chicago Press.

Wilford, J. 1990, *Mars Beckons*, New York: Alfred A. Knopf.

Williamson, R. 1933, *Religious and Cosmic Beliefs of Central Polynesia*, 1, Cambridge: Cambridge Univ. Press.

Wind, E. 1983, *Pagan Mysteries in the Renaissance*, London: Faber & Faber.

Zinner, E. 1947, *The Stars Above Us*, London: Allen & Unwin.

NOTES

3. MYTHOLOGY: NAMING THE IMAGES

1. B. de Sahagún, *Florentine Codex*, Book 7, p. 2.
2. M. Clagett, *Ancient Egyptian Science*, 1:413, 491.
3. J. Pritchard, "Hymn to Ishtar," in *Ancient Near Eastern Texts*, 383.
4. M. Gauquelin, *The Scientific Basis of Astrology*, 82.
5. B. van der Waerden, *Science Awakening II*, 39.
6. J. Pritchard, *Ancient Near Eastern Texts*, 53.
7. Ibid., 55.
8. E. Schafer, *Pacing the Void*, 214.
9. A. Sayce, "Astronomy and Astrology," 199.
10. G. Maspero, *Dawn of Civilization*, 639–40.
11. J. Offord, "The Deity of Crescent Venus in Ancient Western Asia," 200.
12. Ibid.
13. R. Barnes, *Kédang*, 21.
14. C. Singleton, tr. & ed. *Inferno*, BK. XXXIV, lines 34–36.
15. P. Roe, *Cosmic Zygote*, 239–253.
16. D. Tedlock, *Popol Vuh*, 159–60.
17. E. Seler, *Annales de Quauhtitlán*, 364–65.
18. B. de Sahagún, *Florentine Codex*, Book 7, pp. 11–12.
19. A. López Austin, *Sixteenth Century Mexico*, 135–36.
20. J.E.S. Thompson, *Maya Hieroglyphic Writing*, 211, 217.
21. R. Roys, *Book of Chilam Balam*, 101; J. Thompson, *Maya Hieroglyphic Writing*, 219.
22. R. Roys, *Book of Chilam Balam*, 101.
23. Ibid., 122; J.E.S. Thompson, *Maya Hieroglyphic Writing*, 220.
24. Z. Nuttall, "Periodical Adjustment," 497–98.
25. R. Williamson, *Religious and Cosmic Beliefs*, 123.
26. V. Chamberlain, *When Stars Came Down to Earth*, 74.

4. ASTRONOMY: FOLLOWING THE IMAGES

1. J.E.S Thompson, *Maya Hieroglyphic Writing*, 162.
2. Schele and M. Miller, *Blood of Kings*, 301.
3. R.M. Frazer, tr. & ed., *Works & Days*, lines 564–573.
4. A. Tozzer, "Landa's Relación de las Cosas de Yucatan," 27.

5. Ibid., 28, Note 156.
6. B. de Sahagún, *Florentine Codex*, Book 7, pp. 8, 10.
7. P. Huber, "Early Cuneiform Evidence for the Existence of Venus," 123.
8. A. Sachs, "Babylonian Observational Astronomy," 44.
9. S. Langdon, J. Fotheringham, and C. Schoch, *The Venus Tablets of Ammizaduga*, Ch. 10. Brackets added.
10. A. Pannekoek, *History of Astronomy*, 57.
11. Ibid., 55.
12. B. van der Waerden, *Science Awakening II*, 56.

5. ASTROLOGY: BELIEVING IN THE IMAGES

1. W. Fulke, *Book of Meteors*, 84.
2. R. Thompson, *Reports of Magicians and Astrologers*, 2:liii, no. 98 (obv. 1–8).
3. S. J. Tester, *History of Western Astrology*, 185.
4. D. Tedlock, *Popol Vuh*, 174–75.
5. B. van der Waerden, *Science Awakening II*, 58.
6. R. Thompson, *Reports of Magicians and Astrologers*, 2:xci, no. 276, obv. 1; xci, no. 277; lxxxi, no. 260.
7. O. Neugebauer and R. Parker, *Egyptian Astronomical Texts III*, 214–15.
8. A. Oppenheim, "Divination and Celestial Observation," 115.
9. R. Thompson, *Reports of Magicians and Astrologers*, 2:lxx, no. 210, from Asaridu.
10. A. Nathan, *New Republic*, 33.
11. A. Pannekoek, *History of Astronomy*, 88.
12. Ibid., 91.
13. For an excellent discussion of the meaning of *influence*, see J. D. North, "Celestial Influence," 45–107.
14. E. Zinner, *The Stars Above Us*, 39; M. Gauquelin, ibid., ch. 3, Note 4, 85.
15. R. Garland, *The Greek Way of Life*, Ch. 1.
16. R. Gleadow, *Origin of the Zodiac*, 59.
17. D. Ulansey, "Mithraic Mysteries."
18. D. Ulansey, *Origins of the Mithraic Mysteries*, 4.
19. Qtd. in J. Seznec, *Survival of the Pagan Gods*, 44.
20. S. J. Tester, *History of Western Astrology*, 194.
21. St. Bernard, qtd. in E. Garin, *Astronomy in the Renaissance*, 32.
22. E. Wind, *Pagan Mysteries in the Renaissance*, 125.
23. Qtd. in E. Garin, *Astronomy in the Renaissance*, 28, 121.
24. S. J. Tester, *History of Western Astrology*, 133.
25. E. Garin, *Astronomy in the Renaissance*, 8, 122–23.
26. A. Fraknoi, "Your Astrology Defense Kit," 146, 150.

6. TECHNOLOGY: HARNESSING THE IMAGES

1. J. Dobson and S. Brodetsky, tr. & ed. *De Revolutionibus*, 19.
2. Qtd. in S. Drake, *Dialogue*, vii–xix.
3. A. van Helden, tr. and ed., *Siderius Nuncius*, 3–4.
4. R. Latham, tr. & ed., *On the Nature of the Universe*, 142.
5. J. Weinberger, ed., *Great Instauration*, 70–78.
6. A. van Helden, *Siderius Nuncius*, 8.
7. Ibid., 64–65.
8. Ibid., 102.
9. Ibid., 107–8.
10. J. Offord, "The Deity of Crescent Venus in Ancient Western Asia," 197.
11. Bezond, qtd. in ibid., 197.
12. E. Garin, *Astronomy in the Renaissance*, 11.
13. A. Koestler, *Watershed*, 200.
14. S. Drake, *Discoveries*, 175–216; quotation on 176.
15. A. Whatton, *Transit of Venus*, 9.
16. Ibid., 50.
17. R. Proctor, *Transits of Venus*, 20.
18. A. Whatton, *Transit of Venus*, 116–17.
19. R. Proctor, *Transits of Venus*, 29.
20. A. Whatton, *Transit of Venus*, 57.
21. See, e.g., W. Fash and B. Fash, "Scribes, Warriors, and Kings."
22. J. E. S. Thompson, *Commentary*, 77.
23. Qtd. in S. Drake, *Dialogue*, vii.

7. THE IMAGE FOR ITS OWN SAKE

1. A. von Humboldt, *Cosmos*, Vol IV, p. v.
2. J.N. Wilford, *Mars Beckons*, 214.
3. A.N. Whitehead, *Science and the Modern World*, 19–20.
4. D. Bohm, *Wholeness & the Implicate Order*, 7.
5. S. Harding, et al, *"Does Ideology Stop at the Laboratory Door?"* 24.
6. G. Gale, "Anthropic Principle," 154.
7. J. Lovelock, *The Ages of Gaia*, 212.
8. R. Jerome, "Ancient Seas of Mars," 7.
9. S. Weinberg, *The First Three Minutes*, 154.

INDEX

afterlife:
  curiosity about, 48–52, 56–57
  Venus descent myth and, 56
Age of Empiricism, 199
Age of Enlightenment, 183
abau principle, 110, 112, 198
Alexander III (the Great), King of
  Macedon, 141
Alexandria, 141–42, 144
allegories, 162–71, 198
  discovery process in, 168
  transformation of planctary gods
    into figures in, 162–69
Almagest (Ptolemy), 142
*American Ephemeris and Nautical
  Almanac*, 149
Ammizaduga, King of Babylonia,
  142
  hypothetical period of reign of,
    118
  Venus Tablet of, 86–87, 114–27,
    222
Anahita, 36, 48, 61
Anaximenes, 203
Ancient Egyptian Religion (Frankfort),
  128
Andromeda, spiral galaxy in, 210
animals:
  human dominion over, 133
  perception of similarities between
    humans and, 7–8
Antares, 59
anthropic cosmological principle,
  214, 216, 223
Antony, Marc, 141
Anu, 51, 134, 152
Aphrodite, 48, 60–62, 167
apiculture, 15, 72, 74, 76–78
Apocalypse, 212, 216
Apollo man-on-the-moon mission,
  208
Aquarius, 21
  conjunction of Mars, Jupiter, and
    Saturn in, 161–62

Archimedes, 170
architecture, 139
  of Babylonians, 114, 125
  of Mayas, 66, 109, 112–13, 196,
    199, 201
  as means of enlightening humans
    about workings of nature, 196
Aries:
  planetary conjunctions in, 152
  properties of, 146
Aristotle, 14, 84, 127, 139, 168,
  210, 218
  cosmos described by, 153–54
  physics of, 216
Arizona, University of, 217
art:
  celebration of astrology in, 15–16
  discovery process in, 7–8
  in Renaissance, 16, 164–69
  see also specific arts
Assyrians, planets named by, 52
Astrological Lodge, 175
Astrological Principles (Cecco), 161
astrology and astrologers, xi, 10–12,
  15–17, 84–85, 124, 128–77,
  222–23
  in Alexandria, 142, 144
  of Babylonians, 130–32, 139–40,
    144–45, 150–51, 153, 155
  and blood sacrifices, 130–32
  of Chaldeans, 129–30
  and changing images of planets,
    153–69
  of Chinese, 16, 136–38, 150–53
  Christianity in appropriating as-
    pects of, 160
  connection between medicine and,
    143–44
  criticisms of, 155–57, 170–73,
    175–76
  cultural underpinnings of, 16
  death of, 128, 170
  demands on, 135
  of Egyptians, 135, 137

INDEX

INDEX

Galileo (*cont'd*)
  and second death of astrology,
    171–72
  secrecy of, 189
  on separation of actions of humans
    from nature, 17
  telescopes designed by, 182–84
Gemini:
  influence of, 149
  Mars's motion through, 26
Genesis, 44, 133, 218
genethliacal astrology, 129, 170
geometry, 141–42, 149, 154
  definition of, 141
  Ptolemy's use of, 142
  zodiacal signs related through,
    147
George VI, King of England, 175
Ghibellines, 169
Gilgamesh story, 55–56
Giotto, 168
Glashow, Sheldon, 18, 213, 216
God, 208–9
  in determining planetary orbits, 9
  in heliocentric universe, 180
  and origin of universe, 208, 218
  and parting Red Sea, 43
  planets as intermediaries between
    humanity and, 160
  schism between nature and, 17,
    193
  words of, 44
"God in the Machine" (Hawkes), 6
Gordon, R. L., 158
gravitation, concept of, 44
  Newton's discovery of, 7–10, 47,
    154, 156, 168
  planetary influence and, 154–56
Graz, University of, 8
Great Ball Court at Copán, 111,
  199–200
Great Cycles, 101, 107
Great Instauration (Bacon), 183
Great London Fire, 173
Greeks, 53, 56–57, 72, 160–61, 169,
  198, 218, 221
  astrological system of, 139–50,
    154, 156, 160, 170
  astronomers' admiration for, 84

  in charting planets' courses
    through zodiac, 25–27
  creation myths of, 80
  cyclical concept of time of, 91–92
  divining practices of, 129
  earth as center of all things for,
    140–41
  geometry of, 141, 147, 154
  on house system, 148
  influence of Middle Eastern cul-
    tures on, 57
  on influence of moon, 132
  as intellectual lightweights, 14
  on luminosity of planets, 188*n*
  Mayas compared to, 89
  medicine of, 142–44
  Milesian and Ionian philosophers
    of, 203–4
  planets named by, 52
  on power of numbers, 45
  Renaissance and, 166–67, 180
  respect for astronomy among, 10
  on retrograde motion of planets,
    6, 25–27
  scientific theory and, 211–12
  2920-day cycle of, 95
  Venus descent myth of, 56
  Venus dualities of, 62, 165
  Venus named by, 48
  Venus as predictor of winds for,
    80
Gregorian calendar, 107
Gregory, Pope, 107
Group of the Cross at Palenque,
  66–67
Guelphs, 169

halos in astrology, 129–30
Hammurabi, King of Babylonia, 118
Harkhabi, 135
harmonic law, 9–10
Harmony of the Worlds, 10
harvests, rising Venus linked to, 125
Hawaiians, 79–81
Hawkes, Jacquetta, 6
heliacal rise, 28
Henry VI (Shakespeare), 3
hepatoscopy, 84–85
hermaphrodite, origin of word, 60

242

INDEX

INDEX

INDEX

excavating cunieform tablets in, 75, 115
palace economy of, 114
Uxmal, House of the Governor at, 197

van Helden, Albert, 182
Vanity Fair, 175
Vatican Observatory, 187
Velikovsky, Immanuel, 176
Venera spacecraft, 17, 215
Venus, 11–13, 15–18, 34–36, 40–42, 81–82, 143–45, 176, 208
acceptable adjectives for, 5n
allegorical portrayals of, 164–67, 198
androgynous portrayals of, 61
annual horizontal oscillating motion of, 31
apiculture and, 15, 72, 74, 76–78, 94
appreciating predecessors' understanding of, 11
Babylonians on, 15–16, 35, 48, 86–87, 114–27, 222
blood sacrifices to, 69–71, 80
brightness of, 32
Caesar's attachment to, 34, 62
comparisons between earth and, 205–6, 215–17
in conjunctions with Jupiter, 138
in conjunctions with Mars, 138
and continuity of power of rulership, 16, 110–13, 199–201
Copán cult dedicated to, 109–13
costumes of gods associated with, 71–72
crescent observations of, 32, 185–87
criticisms of astrological reckonings of, 173
and curiosity about afterlife, 48–52, 56–57
daylight observation of, 32
in death and resurrection metaphor, 49–51, 53–59, 68–72, 177, 222
as descending bee god, 72, 74, 76–78, 94

disappearance intervals of, 28, 30–33
double-headed sky serpents dedicated to worship of, 42, 50, 111–12
Dresden Codex on, 5, 70–72, 86–87, 96–109, 111, 113n, 115–18, 120–27, 142
dual aspects of, 34, 60–63, 134, 163–67, 169
in Eclipse Table of Dresden Codex, 72, 74, 76–77, 94, 103–6
eight-year cycles of morning and evening star curves of, 31
in evening sky after sunset, 28
fertility associated with, 30, 34, 52–54, 78–79, 112
584-day cycle of, 30–31, 33, 95, 97–98, 100, 103–4, 106–7
587-day period of, 123, 126
future scientific observation of, 216–17
Galileo's observations of, 184–90, 195
as goddess of love, xi, 12, 53–56
grouping cycles of, in fives, 31
Hawaiians on, 79–80
as hero with five faces, 13
Horrocks's observation of, 191–95
human conception associated with, 30, 78–79
humor of, 143
as image of feminine seduction, 36, 61
influence of, 144–45, 202
intervals used to predict future whereabouts of, 123–25
Ixchel's relationship with, 105–6
King Ammizaduga's Tablet on, 86–87, 114–27, 222
Lucifer linked to, 62–63
malevolent omens of, 103
masculine portrayals of, 13, 68–78
Mayas on, 4–6, 8, 15–16, 42, 49–50, 72–78, 94, 101–4, 106, 109–13, 196–97, 199–201
in medieval manuscripts, 40
modern science on, 20
as morning star, 50

INDEX

Venus (*cont'd*)
moving in morning twilight, 28–30, 33
naked-eye observation of, 12, 25–33
in Near East mythology, 47–50
omens linked to risings of, 122–23, 125
as part of triad with sun and moon, 35
in Pawnee creation myth, 80–81
Persians on, 36, 48, 61
phases of, 184–90, 195
as predictor of winds, 80
Ptolemy on, 142
rain omens and, 4–6, 31, 58–59, 100, 111–12, 125
relationship between seasons and, 30–31, 33, 111–12
Renaissance recasting of, 163–69
shifting and flickering of light of, 12–14
shortfall between real and tabular period of, 106–7, 123–24
sky supported by, 41
in space age, 17–18, 203, 205–6, 215–17
special status accorded to, xi, 12, 28, 48
Sumerian worship of, 48, 50–51, 53–59, 62
telescopic probing of, 17, 184–95
terrestrial dwelling place of, 52, 57–58
in time reckoning with moon as principal indicator, 94
in transit across surface of sun, 188, 191–95
260-day cycle of, 78–79, 101–4, 106, 109
warring aspect of, 34, 59–60, 80, 102, 125
see also Inanna; Ishtar
Venus-Moon calendars, 109
of Babylonians, 86–87, 118–19, 123–24, 126
distortions of short-term Venus time in, 104–5, 118, 123–24
dynastic chronologies and, 118–19

and fitting together eclipses and Venus's appearances, 103–4, 107–8
of Mayas, 86–87, 94–98, 103–4, 107–8, 118–19, 123–24, 126
modern calendar compared to, 105
Verdi, Giuseppe, 43
Vernant, Jean-Pierre, 139, 141
Vesper, 48, 62
Viking spacecraft, 215, 217
Virgil, 63, 129
Virgo, 59, 173
von Baeyer, Hans, 183–84, 214
Voyager spacecraft, 204, 207–8

war omens, 34, 59–60, 80, 102, 125
weather, prediction of, 129–30
Weber, Renée, 34
Weinberg, Steven, 14, 218
Whatton, Arundell, 178
Whitehead, Alfred North, 202, 209
Wilford, John Noble, 206
Wind, Edgar, 166
wind, Venus as predictor of, 80
words:
of God, 44
power of, 44–46
Works and Days (Hesiod), 80, 91–92, 215
Worlds in Collision (Velikovsky), 176

Xibalba, 49–50

Yax Pac, King of Maya, 110–11, 113*n*, 199–201
Yoaltecuhtli, 100

Zapotecs, 88
zodiac and zodiac signs, 24–26, 52, 139
Babylonian observational records of planetary positions in, 123
birthstones associated with, 143*n*
and blood types, 175*n*
charting planets' courses through, 25–27, 33
of Chinese, 153
and connection between astrology and medicine, 144

254

INDEX

ABOUT THE AUTHOR

ANTHONY AVENI has pioneered the study of astronomical anthropology and archaeoastronomy. He is the Russell B. Colgate Professor of Astronomy and Anthropology at Colgate University, Hamilton, New York, where he has taught for twenty-five years. He was voted 1982 Professor of the Year by the Council for the Advancement and Support of Education, the United States's highest teaching award, and was in 1991 named one of *Rolling Stone* magazine's top ten university professors. He is the author of *Skywatchers of Ancient Mexico* and *Empires of Time: Calendars, Clocks and Culture*, and has lectured on astronomy for the Learning Channel.

KODANSHA GLOBE

International in scope, this series offers distinguished books that explore the lives, customs, and mindsets of peoples and cultures around the world.

THE GREAT GAME
*The Struggle for Empire
   in Central Asia*
Peter Hopkirk
1-56836-022-3
$15.00

SCENT
*The Mysterious and
   Essential Powers
   of Smell*
Annick Le Guérer
Translated by
   Richard Miller
1-56836-024-X
$13.00

TIGER IN THE BARBED WIRE
*An American in
   Vietnam, 1952–1991*
Howard R. Simpson
Foreword by
   Pierre Salinger
1-56836-025-8
$16.00

ZEN IN AMERICA
*Five Teachers and the
   Search for an
   American Buddhism*
Helen Tworkov
Foreword by
   Natalie Goldberg
New epilogue by
   the author
1-56836-030-4
$15.00

THE HEART OF THE SKY
Travels Among the Maya
Peter Canby
New introduction by
   the author on
   the Maya peasant
   rebellion
1-56836-026-6
$13.00

**CONVERSING WI
   PLANETS**
*How Science a
   Invented th*
Anthony Aveni
1-56836-021-5
$14.00

**SONS OF THE YE
   EMPEROR**
*The History of t..
   Chinese Diaspora*
Lynn Pan
New introduction by
   the author
1-56836-032-0
$15.00

**THE WORLD OF THE
   SHINING PRINCE**
*Court Life in Ancient
   Japan*
Ivan Morris
Introduction by
   Barbara Ruch
1-56836-029-0
$15.00

BLOODTIES
*Nature, Culture and the
   Hunt*
Ted Kerasote
1-56836-027-4
$13.00

AS WE SAW THEM
*The First Japanese
   Embassy to the
   United States*
Masao Miyoshi
1-56836-028-2
$13.00

WARRIOR OF ZEN
*The Diamond-hard
   Wisdom Mind of
   Suzuki Shôsan*
Edited, translated
   and introduced by
   Arthur Braverman
1-56836-031-2
$10.00

To order, contact your local bookseller or call 1-800-788-6262 (mention code G1). For information on future titles, please contact the Kodansha Editorial Department at Kodansha America, Inc., 114 Fifth Avenue, New York, NY 10011.